CW00524432

TEXTS IN COMPUTER SCIENCE

Editors
David Gries
Fred B. Schneider

Springer
New York
Berlin
Heidelberg
Barcelona
Hong Kong
London
Milan
Paris
Singapore
Tokyo

TEXTS IN COMPUTER SCIENCE

(continued after index)

Colin Stirling

MODAL AND TEMPORAL PROPERTIES OF PROCESSES

With 45 Illustrations

Colin Stirling
Division of Informatics
University of Edinburgh
Edinburgh EH9 3JZ UK
cps@dcs.ed.ac.uk

Series Editors

David Gries
Department of Computer Science
415 Boyd Studies Research Center
The University of Georgia
Athens, GA 30605, USA

Fred B. Schneider
Department of Computer Science
Upson Hall
Cornell University
Ithaca, NY 14853-7501, USA

Library of Congress Cataloging-in-Publication Data
Stirling, Colin P.
Modal and temporal properties of processes / Colin Stirling.
p. cm. – (Texts in computer science)
Includes bibliographical references and index.
ISBN 0-387-98717-7 (alk. paper)
1. Computer logic. 2. Parallel processing (Electronic computers) I. Title. II. Series.
QA76.9.L63 S75 2001
004′.35–dc21 00–067924

Printed on acid-free paper.

Production managed by Timothy Taylor; manufacturing supervised by Erica Bresler.
Typeset pages prepared using the author's LaTeX 2_ε files by The Bartlett Press, Inc., Marietta, GA.
Printed and bound by Hamilton Printing Co., Rensselaer, NY.
Printed in the United States of America.

9 8 7 6 5 4 3 2 1

ISBN 0-387-98717-7 SPIN 10708024

Springer-Verlag New York Berlin Heidelberg

A member of BertelsmannSpringer Science+Business Media GmbH

To Sarah and Susie

Preface

In this book we examine modal and temporal logics for processes. First, we introduce concurrent processes as terms of an algebraic language comprising a few basic operators. Their behaviours are described using transitions. Families of transitions can be arranged as labelled graphs, concrete summaries of the behaviour of processes. Various combinations of processes and their resulting behaviour, as determined by the transition rules, are reviewed. Next, simple modal logics are introduced for describing the capabilities of processes.

An important discussion point occurs when two processes may be deemed to have the same behaviour. Such an abstraction can be presented by defining an appropriate behavioural equivalence between processes. A more abstract approach is to consider equivalence in terms of having the same pertinent properties. There is special emphasis with bisimulation equivalence, since the discriminating power of modal logic is tied to it.

More generally, practitioners have found it useful to be able to express temporal properties of concurrent systems, especially liveness and safety properties. A safety property amounts to "nothing bad ever happens," whereas a liveness property expresses "something good does eventually happen." The crucial safety property of a mutual exclusion algorithm is that no two processes are ever in their critical sections concurrently. And an important liveness property is that, whenever a process requests execution of its critical section, then eventually it is granted. Cyclic properties of systems are also salient: for instance, part of a specification of a scheduler is that it must continually perform a particular sequence of actions. A logic expressing temporal notions provides a framework for the precise formalisation of such specifications.

Formulas of the modal logic are not rich enough to express such temporal properties, so an extra, fixed point operator, is added. The result is a very expressive temporal logic, modal mu-calculus. However, it is also very important to be able to verify that an agent has or does not have a particular property.

The text aims to be reasonably introductory, so that parts of the book could be used at undergraduate level, as well as at more advanced levels. I have used the material in this way at Edinburgh. The extensive use of games for both equivalence and model checking is partly pedagogical, since they are so conceptually clear.

Parts of the book have been presented previously at various summerschools over the years, and I wish to thank all the organisers for allowing me to present this material. I should also like to thank current and previous colleagues at Edinburgh for building such an intellectually stimulating environment to work in. In particular, I wish to thank Julian Bradfield (a pioneer of infinite state model checking and who allowed me to use his TeX tree constructor for building derivation trees), Olaf Burkart, Kim Larsen (who introduced me to modal mu-calculus), Robin Milner (from whom I learnt about process calculus and bisimulation equivalence), Perdita Stevens and David Walker.

Colin Stirling
Edinburgh, United Kingdom

Contents

List of Figures

1

Processes

In this chapter, processes are introduced as expressions of a simple language built from a few basic operators. The behaviour of a process E is characterised by transitions of the form $E \xrightarrow{a} F$, that E may become F by performing the action a. Structural rules prescribe behaviour, since the transitions of a compound process are determined by those of its components. Concrete pictorial summaries of behaviour are presented as labelled graphs, which are collections of transitions. We review various combinations of processes and their resulting behaviour.

1.1 First examples

A simple process is a clock that perpetually ticks.

$$\texttt{Cl} \stackrel{\text{def}}{=} \texttt{tick.Cl}$$

Names of actions such as `tick` are in lower case, whereas names of processes such as `Cl` have an initial capital letter. A process definition ties a process name to a process expression. In this case, `Cl` is attached to `tick.Cl`, where both occurrences of `Cl` name the same process. The defining expression for `Cl` invokes a prefix operator . that builds the process $a.E$ from the action a and the process E.

Behaviour of processes is captured by transitions $E \stackrel{a}{\longrightarrow} F$, that E may evolve to F by performing or accepting the action a. The behaviour of `Cl` is elementary, since it can only perform `tick` and in so doing becomes `Cl` again. This is a consequence of the rules for deriving transitions. First is the axiom for the prefix operator.

$$\boxed{\mathrm{R}(.) \quad a.E \stackrel{a}{\longrightarrow} E}$$

A process $a.E$ performs the action a and becomes E. An instance of this axiom is the transition $\mathtt{tick.Cl} \stackrel{\mathtt{tick}}{\longrightarrow} \mathtt{Cl}$. The next transition rule refers to the operator $\stackrel{\mathrm{def}}{=}$, and is presented with the desired conclusion uppermost.

$$\boxed{\mathrm{R}(\stackrel{\mathrm{def}}{=}) \quad \frac{P \stackrel{a}{\longrightarrow} F}{E \stackrel{a}{\longrightarrow} F} \quad P \stackrel{\mathrm{def}}{=} E}$$

If the transition $E \stackrel{a}{\longrightarrow} F$ is derivable and $P \stackrel{\mathrm{def}}{=} E$, then $P \stackrel{a}{\longrightarrow} F$ is also derivable. Goal-directed transition rules are used because we are interested in discovering the available transitions of a process. There is a single transition for the clock, $\mathtt{Cl} \stackrel{\mathtt{tick}}{\longrightarrow} \mathtt{Cl}$. Suppose our goal is to derive a transition $\mathtt{Cl} \stackrel{a}{\longrightarrow} E$. Because the only applicable rule is $\mathrm{R}(\stackrel{\mathrm{def}}{=})$, the goal reduces to the subgoal $\mathtt{tick.Cl} \stackrel{a}{\longrightarrow} E$, and the only possibility for deriving this subgoal is an application of R(.), in which case a is `tick` and E is `Cl`.

The behaviour of `Cl` is represented graphically in Figure 1.1. Ingredients of this behaviour graph (known as a "transition system") are process expressions and binary transition relations between them. Each vertex is a process expression, and one of the vertices is the initial vertex `Cl`. Each derivable transition of a vertex is depicted. Transition systems abstract from the derivations of transitions.

An unsophisticated vending machine `Ven` is defined in Figure 1.2. The definition of `Ven` employs the binary choice operator $+$ (which has wider scope than the prefix operator) from Milner's CCS, Calculus of Communicating Systems [42, 44]. Initially `Ven` may accept a 2p or 1p coin, and then a button `big` or `little` may be depressed depending on the coin deposited, and finally after an item is

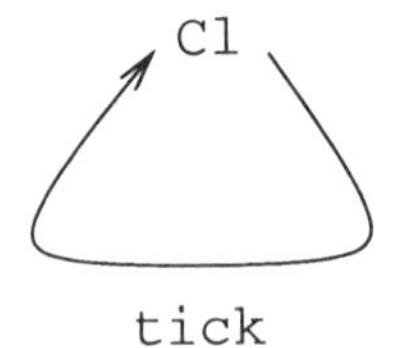

FIGURE 1.1. The transition graph for `Cl`

$$\begin{array}{lcl} \texttt{Ven} & \stackrel{\text{def}}{=} & \texttt{2p.Ven}_\texttt{b} + \texttt{1p.Ven}_\texttt{l} \\ \texttt{Ven}_\texttt{b} & \stackrel{\text{def}}{=} & \texttt{big.collect}_\texttt{b}\texttt{.Ven} \\ \texttt{Ven}_\texttt{l} & \stackrel{\text{def}}{=} & \texttt{little.collect}_\texttt{l}\texttt{.Ven} \end{array}$$

FIGURE 1.2. A vending machine

collected the process reverts to its initial state. There are two transition rules for $+$.

$$\text{R}(+) \quad \frac{E_1 + E_2 \stackrel{a}{\longrightarrow} F}{E_1 \stackrel{a}{\longrightarrow} F} \qquad \frac{E_1 + E_2 \stackrel{a}{\longrightarrow} F}{E_2 \stackrel{a}{\longrightarrow} F}$$

The derivation of the transition $\texttt{Ven} \stackrel{\texttt{2p}}{\longrightarrow} \texttt{Ven}_\texttt{b}$ is as follows.

$$\cfrac{\texttt{Ven} \stackrel{\texttt{2p}}{\longrightarrow} \texttt{Ven}_\texttt{b}}{\cfrac{\texttt{2p.Ven}_\texttt{b} + \texttt{1p.Ven}_\texttt{l} \stackrel{\texttt{2p}}{\longrightarrow} \texttt{Ven}_\texttt{b}}{\texttt{2p.Ven}_\texttt{b} \stackrel{\texttt{2p}}{\longrightarrow} \texttt{Ven}_\texttt{b}}}$$

The goal reduces to the subgoal beneath it as a result of an application of $\text{R}(\stackrel{\text{def}}{=})$, which in turn reduces to the axiom instance via an application of the first of the $\text{R}(+)$ rules. When presenting proofs of transitions, side conditions in the application of a rule, such as $\text{R}(\stackrel{\text{def}}{=})$, are omitted. Figure 1.3 pictures the transition system for Ven.

A transition $E \stackrel{a}{\longrightarrow} F$ is an assertion derivable from the rules for transitions. To discover the transitions of E, it suffices to examine its main combinator and the transitions of its components. There is an analogy with rules for expression evaluation. To evaluate $(3 \times 2) + 4$ it suffices to evaluate the components 3×2 and 4, and then sum their values. Such families of rules give rise to a structural

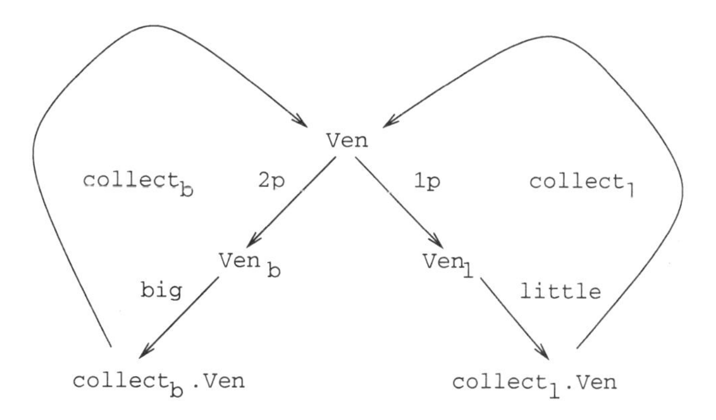

FIGURE 1.3. The transition graph for Ven

$$\begin{aligned} \mathtt{Ct}_0 &\stackrel{\text{def}}{=} \mathtt{up}.\mathtt{Ct}_1 + \mathtt{round}.\mathtt{Ct}_0 \\ \mathtt{Ct}_{i+1} &\stackrel{\text{def}}{=} \mathtt{up}.\mathtt{Ct}_{i+2} + \mathtt{down}.\mathtt{Ct}_i \end{aligned}$$

FIGURE 1.4. A family of counters

operational semantics, as pioneered by Plotkin [49]. However, whereas the essence of an expression is to be evaluated, the essence of a process is to act.

Families of processes can be defined using indexing. A simple case is the set of counters $\{\mathtt{Ct}_i \ : \ i \in \mathbb{N}\}$ of Figure 1.4. The counter $\mathtt{Ct}_3$ can increase to $\mathtt{Ct}_4$ by performing up or decrease to $\mathtt{Ct}_2$ by performing down. The derivation of the transition $\mathtt{Ct}_3 \xrightarrow{\mathtt{up}} \mathtt{Ct}_4$ is as follows.

$$\cfrac{\cfrac{\mathtt{Ct}_3 \xrightarrow{\mathtt{up}} \mathtt{Ct}_4}{\mathtt{up}.\mathtt{Ct}_4 + \mathtt{down}.\mathtt{Ct}_2 \xrightarrow{\mathtt{up}} \mathtt{Ct}_4}}{\mathtt{up}.\mathtt{Ct}_4 \xrightarrow{\mathtt{up}} \mathtt{Ct}_4}$$

The rule $\mathrm{R}(\stackrel{\text{def}}{=})$ is here applied to the instance $\mathtt{Ct}_3 \stackrel{\text{def}}{=} \mathtt{up}.\mathtt{Ct}_4 + \mathtt{down}.\mathtt{Ct}_2$. Each member $\mathtt{Ct}_\mathtt{i}$ determines the same transition graph of Figure 1.5 which contains an infinite number of vertices. This graph is "infinite state" because the behaviour of $\mathtt{Ct}_\mathtt{i}$ may progress through any of the processes $\mathtt{Ct}_\mathtt{j}$, in contrast to the finite state graphs of Figures 1.1 and 1.3.

The operator $+$ can be extended to indexed families $\sum\{E_i \ : \ i \in I\}$ where I is a set of indices. $E_1 + E_2$ abbreviates $\sum\{E_i \ : \ i \in \{1,2\}\}$. Indexed sum may be coupled with indexing of actions. An example is a register storing numbers, represented as a family $\{\mathtt{Reg}'_i \ : \ i \in \mathbb{N}\}$.

$$\mathtt{Reg}'_\mathtt{i} \stackrel{\text{def}}{=} \mathtt{read}_\mathtt{i}.\mathtt{Reg}'_\mathtt{i} + \sum\{\mathtt{write}_j.\mathtt{Reg}'_j \ : \ j \in \mathbb{N}\}$$

The act of reading the content of the register when i is stored is $\mathtt{read}_\mathtt{i}$, whereas $\mathtt{write}_\mathtt{j}$ is the action that updates its value to j. The single transition rule for $\sum$ generalises the rules for $+$.

$$\mathrm{R}(\textstyle\sum) \quad \cfrac{\sum\{E_i \ : \ i \in I\} \xrightarrow{a} F}{E_j \xrightarrow{a} F} \quad j \in I$$

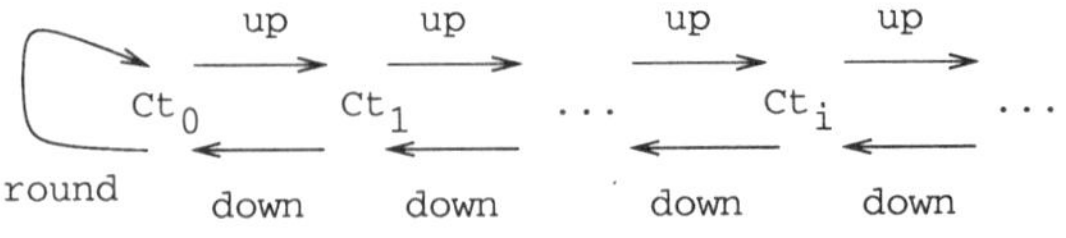

FIGURE 1.5. The transition graph for $\mathtt{Ct}_i$

Consequently, $\mathtt{Reg}'_\mathtt{i}$ is able to carry out any $\mathtt{write}_\mathtt{j}$ (and thereby changes to $\mathtt{Reg}'_\mathtt{j}$) as well as $\mathtt{read}_\mathtt{i}$ (and then remains unchanged). A special case is when the indexing set I is empty. By the rule $\mathrm{R}(\sum)$, this process has no transitions, since the subgoal can never be fulfilled. In CCS the nil process $\sum\{E_i : i \in \emptyset\}$ is abbreviated to 0 (and to STOP in Hoare's CSP, Communicating Sequential Processes [31]).

Actions can be viewed as ports or channels, means by which processes can interact. It is then also important to consider the passage of data between processes along these channels, or through these ports. In CCS, input of data at a port named a is represented by the prefix $a(x).E$, where $a(x)$ binds free occurrences of x in E. (In CSP $a(x)$ is written $a?x$.) The port label a no longer names a single action, instead it represents the set $\{a(v) : v \in D\}$ where D is the appropriate family of data values. The transition axiom for this prefix input form is

$$\boxed{\mathrm{R(in)}\quad a(x).E \xrightarrow{a(v)} E\{v/x\} \ \text{ if } \ v \in D}$$

where $E\{v/x\}$ is the process term that results from replacing all free occurrences of x in E with v[1]. Output at a port named a is represented in CCS by the prefix $\overline{a}(e).E$ where e is a data expression. The overbar $^-$ symbolises output at the named port. (In CSP $\overline{a}(e)$ is written $a!e$.) The transition rule for output depends on extra machinery for expression evaluation. Assume that $\mathrm{Val}(e)$ is the data value in D (if there is one) to which e evaluates.

$$\boxed{\mathrm{R(out)}\quad \overline{a}(e).E \xrightarrow{\overline{a}(v)} E \ \text{ if } \mathrm{Val}(e) = v}$$

The asymmetry between input and output is illustrated by the following process that copies a value from in and then sends it through out.

$$\mathtt{Cop} \stackrel{\mathrm{def}}{=} \mathtt{in}(x).\overline{\mathtt{out}}(x).\mathtt{Cop}$$

Below is a derivation of the transition $\mathtt{Cop} \xrightarrow{\mathtt{in}(v)} \overline{\mathtt{out}}(v).\mathtt{Cop}$ for $v \in D$.

$$\frac{\mathtt{Cop} \xrightarrow{\mathtt{in}(v)} \overline{\mathtt{out}}(v).\mathtt{Cop}}{\mathtt{in}(x).\overline{\mathtt{out}}(x).\mathtt{Cop} \xrightarrow{\mathtt{in}(v)} \overline{\mathtt{out}}(v).\mathtt{Cop}}$$

The subgoal is an instance of R(in), as $(\overline{\mathtt{out}}(x).\mathtt{Cop})\{v/x\}$ is $\overline{\mathtt{out}}(v).\mathtt{Cop}$[2], and so the goal follows by an application of $\mathrm{R}(\stackrel{\mathrm{def}}{=})$. The process $\overline{\mathtt{out}}(v).\mathtt{Cop}$ has only one transition $\overline{\mathtt{out}}(v).\mathtt{Cop} \xrightarrow{\overline{\mathtt{out}}(v)} \mathtt{Cop}$ that is an instance of R(out), since we assume that $\mathrm{Val}(v)$ is v. Whenever Cop inputs a value at in, it immediately disgorges it through out. The size of the transition graph for Cop depends on the size of the data domain D, and is finite when D is a finite set.

[1] The process $a(x).E$ can be viewed as an abbreviation of the process $\sum\{a_v.E\{v/x\} : v \in D\}$, writing a_v instead of $a(v)$.

[2] Cop contains no free variables because $\mathtt{in}(x)$ binds x, and so $(\overline{\mathtt{out}}(x).\mathtt{Cop})\{v/x\}$ equals $\overline{\mathtt{out}}(v).(\mathtt{Cop}\{v/x\})$ because x is free in $\overline{\mathtt{out}}(x)$, and $(\mathtt{Cop}\{v/x\})$ is Cop.

Example 1 $\mathtt{Cop}_1 \stackrel{\text{def}}{=} \mathtt{in}(x).\mathtt{in}(x).\overline{\mathtt{out}}(x).\mathtt{Cop}_1$ is a different copier. It takes in two data values at in, discarding the first but sending out the second. $\mathtt{Cop}_1$ has initial transition $\mathtt{Cop}_1 \xrightarrow{\mathtt{in}(v)} \mathtt{in}(x).\overline{\mathtt{out}}(x).\mathtt{Cop}_1$ for $v \in D$.

Input actions and indexing can be mingled, as in the following redescription of the family of registers, where both i and x have type $\mathbb{N}$.

$$\mathtt{Reg}_\mathtt{i} \stackrel{\text{def}}{=} \overline{\mathtt{read}}(i).\mathtt{Reg}_\mathtt{i} + \mathtt{write}(x).\mathtt{Reg}_x$$

$\mathtt{Reg}_\mathtt{i}$ can output the value i at the port read, or instead it can be updated by being written to at write. Below is the derivation of $\mathtt{Reg}_5 \xrightarrow{\mathtt{write}(3)} \mathtt{Reg}_3$.

$$\frac{\dfrac{\mathtt{Reg}_5 \xrightarrow{\mathtt{write}(3)} \mathtt{Reg}_3}{\overline{\mathtt{read}}(5).\mathtt{Reg}_5 + \mathtt{write}(x).\mathtt{Reg}_x \xrightarrow{\mathtt{write}(3)} \mathtt{Reg}_3}}{\mathtt{write}(x).\mathtt{Reg}_x \xrightarrow{\mathtt{write}(3)} \mathtt{Reg}_3}$$

The variable x in $\mathtt{write}(x)$ binds the free occurrence of x in $\mathtt{Reg}_x$. An index can also be presented explicitly as a parameter.

Example 2 The multiple copier $\mathtt{Cop}'$ uses the parameterised subprocess $\mathtt{Cop}(n, x)$, where n ranges over $\mathbb{N}$ and x over texts.

$$\begin{aligned} \mathtt{Cop}' &\stackrel{\text{def}}{=} \mathtt{no}(n).\mathtt{in}(x).\mathtt{Cop}(n, x) \\ \mathtt{Cop}(0, x) &\stackrel{\text{def}}{=} \overline{\mathtt{out}}(x).\mathtt{Cop}' \\ \mathtt{Cop}(i + 1, x) &\stackrel{\text{def}}{=} \overline{\mathtt{out}}(x).\mathtt{Cop}(i, x) \end{aligned}$$

The initial transition of $\mathtt{Cop}'$ determines the number of extra copies of a manuscript, for instance $\mathtt{Cop}' \xrightarrow{\mathtt{no}(4)} \mathtt{in}(x).\mathtt{Cop}(4, x)$. The next transition settles on the text, $\mathtt{in}(x).\mathtt{Cop}(4, x) \xrightarrow{\mathtt{in}(v)} \mathtt{Cop}(4, v)$. Then before reverting to the initial state, five copies of v are transmitted through the port out.

Data expressions may involve operations on values, as in the following example, where x and y range over a space of messages.

$$\mathtt{App} \stackrel{\text{def}}{=} \mathtt{in}(x).\mathtt{in}(y).\overline{\mathtt{out}}(x^\wedge y).\mathtt{App}$$

App receives two messages m and n on in and transmits their concatenation $m^\wedge n$ on out. We shall assume different expression types, such as boolean expressions. An example is that $\text{Val}(\mathit{even}(i)) = \text{true}$ if i is an even integer and is false otherwise. This allows us to use conditionals in the definition of a process as exemplified by S that sieves odd and even numbers.

$$\mathtt{S} \stackrel{\text{def}}{=} \mathtt{in}(x).\textbf{if } \mathit{even}(x) \textbf{ then } \overline{\mathtt{out}}_\mathtt{e}(x).\mathtt{S} \textbf{ else } \overline{\mathtt{out}}_\mathtt{o}(x).\mathtt{S}$$

Below are the transition rules for the conditional.

$$\text{R(if 1)} \quad \frac{\textbf{if } b \textbf{ then } E_1 \textbf{ else } E_2 \xrightarrow{a} E'}{E_1 \xrightarrow{a} E'} \quad \text{Val}(b) = \text{true}$$

$$\text{R(if 2)} \quad \frac{\textbf{if } b \textbf{ then } E_1 \textbf{ else } E_2 \xrightarrow{a} E'}{E_2 \xrightarrow{a} E'} \quad \text{Val}(b) = \text{false}$$

S initially receives a numerical value through the port in. For instance, $\texttt{S} \xrightarrow{\texttt{in}(55)}$ **if** *even*(55) **then** $\overline{\texttt{out}_e}(55).\texttt{S}$ **else** $\overline{\texttt{out}_o}(55).\texttt{S}$. It then outputs through $\texttt{out}_e$ if the received value is even, or through $\texttt{out}_o$ otherwise. In this example, **if** *even*(55) **then** $\overline{\texttt{out}_e}(55).\texttt{S}$ **else** $\overline{\texttt{out}_o}(55).\texttt{S} \xrightarrow{\overline{\texttt{out}_o}(55)} \texttt{S}$.

Example 3 Consider the following family of processes for $i \geq 1$.

$$\texttt{T}(i) \stackrel{\text{def}}{=} \textbf{if } even(i) \textbf{ then } \overline{\texttt{out}}(i).\texttt{T}(i/2) \textbf{ else } \overline{\texttt{out}}(i).\texttt{T}((3i+1)/2)$$

So T(5) performs the sequence of transitions

$$\texttt{T}(5) \xrightarrow{\overline{\texttt{out}}(5)} \texttt{T}(8) \xrightarrow{\overline{\texttt{out}}(8)} \texttt{T}(4) \xrightarrow{\overline{\texttt{out}}(4)} \texttt{T}(2)$$

and then cycles through the transitions $\texttt{T}(2) \xrightarrow{\overline{\texttt{out}}(2)} \texttt{T}(1) \xrightarrow{\overline{\texttt{out}}(1)} \texttt{T}(2)$.

Exercises

1. Draw the transition graphs for the following clocks.
   a. $\texttt{Cl}_1 \stackrel{\text{def}}{=} \texttt{tick.tock.Cl}_1$
   b. $\texttt{Cl}_2 \stackrel{\text{def}}{=} \texttt{tick.tick.Cl}_2$
   c. $\texttt{Cl}_3 \stackrel{\text{def}}{=} \texttt{tick.Cl}$
   d. tick.0
2. Show that there are two derivations of the transition $\texttt{Cl}_4 \xrightarrow{\texttt{tick}} \texttt{Cl}_4$ when $\texttt{Cl}_4 \stackrel{\text{def}}{=} \texttt{tick.Cl}_4 + \texttt{tick.Cl}_4$. Draw the transition graph for $\texttt{Cl}_4$.
3. Contrast the behaviour of $\texttt{Cl}_5 \stackrel{\text{def}}{=} \texttt{tick.Cl}_5 + \texttt{tick.0}$ with that of Cl by drawing their transition graphs.
4. Define a more rational vending machine than Ven that allows the big button to be pressed if two 1p coins are entered, and the little button to be depressed twice after a 2p coin is deposited.
5. Assume that the space of values consists of two elements, 0 and 1. Draw transition graphs for the following three copiers Cop, $\texttt{Cop}_1$ and $\texttt{Cop}_2$ where $\texttt{Cop}_2 \stackrel{\text{def}}{=} \texttt{in}(x).\overline{\texttt{out}}(x).\overline{\texttt{out}}(x).\texttt{Cop}_2$.
6. Draw transition graphs of T(31) and T(17), where T(i) is defined in Example 3.

7. For any processes E, F and G, show that the transition graphs for $E + F$ and $F + E$ are isomorphic, and that the transition graph for $(E + F) + G$ is isomorphic to that of $E + (F + G)$.
8. From Walker [60]. Define a process Change that describes a change-making machine with one input port and one output port, that is capable initially of accepting either a 20p or a 10p coin, and that can then dispense any sequence of 1p, 2p, 5p and 10p coins, the sum of whose values is equal to that of the coin accepted, before returning to its initial state.

1.2 Concurrent interaction

A compelling feature of process theory is modelling of concurrent interaction. A prevalent approach is to appeal to handshake communication as primitive. At any one time, only two processes may communicate at a port or along a channel. In CCS, the resultant communication is a *completed* internal action. Each incomplete, or observable, action a has a partner $\overline{a}$, its co-action. Moreover, the action $\overline{\overline{a}}$ is a, which means that a is also the co-action of $\overline{a}$. The partner of a parameterised action $\texttt{in}(v)$ is $\overline{\texttt{in}}(v)$. Simultaneously performing an action and its co-action produces the internal action τ, which is a complete action that does not have a partner.

Concurrent composition of E and F is expressed as $E \mid F$. Below is the crucial transition rule for $\mid$ that conveys communication.

$$\text{R}(\mid \text{com}) \quad \frac{E \mid F \xrightarrow{\tau} E' \mid F'}{E \xrightarrow{a} E' \quad F \xrightarrow{\overline{a}} F'}$$

If E can carry out an action and become E', and F can carry out its co-action and become F' then $E \mid F$ can perform the completed internal action τ and become $E' \mid F'$. Consider a potential user of the copier Cop of the previous section, who first writes a file before sending it through the port in.

$$\begin{aligned} \texttt{User} &\stackrel{\text{def}}{=} \texttt{write}(x).\texttt{User}_x \\ \texttt{User}_v &\stackrel{\text{def}}{=} \overline{\texttt{in}}(v).\texttt{User} \end{aligned}$$

As soon as User has written the file v, it becomes the process $\texttt{User}_v$ that can communicate with Cop at the port in. Rule R($\mid$ com) is used in the following derivation[3] of the transition $\mathtt{Cop} \mid \mathtt{User_v} \xrightarrow{\tau} \overline{\mathtt{out}}(v).\mathtt{Cop} \mid \mathtt{User}$.

$$\dfrac{\mathtt{Cop} \mid \mathtt{User_v} \xrightarrow{\tau} \overline{\mathtt{out}}(v).\mathtt{Cop} \mid \mathtt{User}}{\dfrac{\mathtt{Cop} \xrightarrow{\mathtt{in}(v)} \overline{\mathtt{out}}(v).\mathtt{Cop}}{\mathtt{in}(x).\overline{\mathtt{out}}(x).\mathtt{Cop} \xrightarrow{\mathtt{in}(v)} \overline{\mathtt{out}}(v).\mathtt{Cop}} \quad \dfrac{\mathtt{User_v} \xrightarrow{\overline{\mathtt{in}}(v)} \mathtt{User}}{\overline{\mathtt{in}}(v).\mathtt{User} \xrightarrow{\overline{\mathtt{in}}(v)} \mathtt{User}}}$$

The goal transition is the resultant communication at $\mathtt{in}$. Through this communication, the value v is sent from the user to the copier because $\mathtt{User_v}$ performs the output $\overline{\mathtt{in}}(v)$ and $\mathtt{Cop}$ performs the input $\mathtt{in}(v)$, where they agree on the value v. Data is thereby passed from one process to another. When the actions a and $\overline{a}$ do not involve values, the resulting communication is a synchronization.

Several users can share the copying resource. $\mathtt{Cop} \mid (\mathtt{User_{v1}} \mid \mathtt{User_{v2}})$ involves two users, but only one at a time is allowed to employ it. So, other transition rules for $\mid$ are needed, permitting components to proceed without communicating.

$$\mathrm{R}(\mid) \quad \dfrac{E \mid F \xrightarrow{a} E' \mid F}{E \xrightarrow{a} E'} \qquad \dfrac{E \mid F \xrightarrow{a} E \mid F'}{F \xrightarrow{a} F'}$$

In the first of these rules, the process F does not contribute to the action a that E performs. Below is a sample derivation.

$$\dfrac{\mathtt{Cop} \mid (\mathtt{User_{v1}} \mid \mathtt{User_{v2}}) \xrightarrow{\tau} \overline{\mathtt{out}}(v1).\mathtt{Cop} \mid (\mathtt{User} \mid \mathtt{User_{v2}})}{\dfrac{\mathtt{Cop} \xrightarrow{\mathtt{in}(v1)} \overline{\mathtt{out}}(v1).\mathtt{Cop}}{\mathtt{in}(x).\overline{\mathtt{out}}(x).\mathtt{Cop} \xrightarrow{\mathtt{in}(v1)} \overline{\mathtt{out}}(v1).\mathtt{Cop}} \quad \dfrac{\mathtt{User_{v1}} \mid \mathtt{User_{v2}} \xrightarrow{\overline{\mathtt{in}}(v1)} \mathtt{User} \mid \mathtt{User_{v2}}}{\dfrac{\mathtt{User_{v1}} \xrightarrow{\overline{\mathtt{in}}(v1)} \mathtt{User}}{\overline{\mathtt{in}}(v1).\mathtt{User} \xrightarrow{\overline{\mathtt{in}}(v1)} \mathtt{User}}}}$$

The goal transition reflects a communication between $\mathtt{Cop}$ and $\mathtt{User_{v1}}$, meaning $\mathtt{User_{v2}}$ is not a contributor. $\mathtt{Cop} \mid (\mathtt{User_{v1}} \mid \mathtt{User_{v2}})$ is not forced to engage in communication. Instead, it may carry out an input action $\mathtt{in}(v)$, or an output action $\overline{\mathtt{in}}(v1)$ or $\overline{\mathtt{in}}(v2)$.

$$\mathtt{Cop} \mid (\mathtt{User_{v1}} \mid \mathtt{User_{v2}}) \xrightarrow{\mathtt{in}(v)} \overline{\mathtt{out}}(v).\mathtt{Cop} \mid (\mathtt{User_{v1}} \mid \mathtt{User_{v2}})$$

$$\mathtt{Cop} \mid (\mathtt{User_{v1}} \mid \mathtt{User_{v2}}) \xrightarrow{\overline{\mathtt{in}}(v1)} \mathtt{Cop} \mid (\mathtt{User} \mid \mathtt{User_{v2}})$$

$$\mathtt{Cop} \mid (\mathtt{User_{v1}} \mid \mathtt{User_{v2}}) \xrightarrow{\overline{\mathtt{in}}(v2)} \mathtt{Cop} \mid (\mathtt{User_{v1}} \mid \mathtt{User})$$

[3] We assume that $\mid$ has greater scope than other process operators. The process $\overline{\mathtt{out}}(v).\mathtt{Cop} \mid \mathtt{User}$ is therefore the parallel composition of $\overline{\mathtt{out}}(v).\mathtt{Cop}$ and $\mathtt{User}$.

The second of these transitions is derived using two applications of R(|).

$$\dfrac{\dfrac{\dfrac{\mathtt{User}_{v1} \xrightarrow{\overline{\mathtt{in}}(v1)} \mathtt{User}}{\mathtt{User}_{v1} \mid \mathtt{User}_{v2} \xrightarrow{\overline{\mathtt{in}}(v1)} \mathtt{User} \mid \mathtt{User}_{v2}}}{\mathtt{Cop} \mid (\mathtt{User}_{v1} \mid \mathtt{User}_{v2}) \xrightarrow{\overline{\mathtt{in}}(v1)} \mathtt{Cop} \mid (\mathtt{User} \mid \mathtt{User}_{v2})}}{}$$

$$\dfrac{\overline{\mathtt{in}}(v1).\mathtt{User} \xrightarrow{\overline{\mathtt{in}}(v1)} \mathtt{User}}{}$$

The behaviour of the users sharing the copier is not impaired by the order of parallel subcomponents, or by placement of brackets. Both processes $(\mathtt{Cop} \mid \mathtt{User}_{v1}) \mid \mathtt{User}_{v2}$ and $\mathtt{User}_{v1} \mid (\mathtt{Cop} \mid \mathtt{User}_{v2})$ have the same capabilities as $\mathtt{Cop} \mid (\mathtt{User}_{v1} \mid \mathtt{User}_{v2})$. These three process expressions have isomorphic transition graphs, and therefore in the sequel we omit brackets between multiple concurrent processes[4].

The parallel operator is expressively powerful. It can be used to describe infinite state systems without invoking infinite indices or value spaces. A simple example is the following counter Cnt.

$$\mathtt{Cnt} \stackrel{\text{def}}{=} \mathtt{up}.(\mathtt{Cnt} \mid \mathtt{down}.0)$$

Cnt can perform up and become Cnt | down.0 that can perform down, or a further up and become Cnt | down.0 | down.0, and so on.

Figure 1.6 offers an alternative pictorial representation of the copier Cop and user User. Such diagrams are called "flow graphs" by Milner [44] (and should be distinguished from transition graphs). A flow graph summarizes the potential movement of information flowing into and out of ports, and also exhibits the ports through which a process is, in principle, willing to communicate. In the case of User, the incoming arrow to the port labelled write represents input, whereas the outgoing arrow from $\overline{\mathtt{in}}$ symbolises output. Figure 1.7 shows the flow graph for Cop | User with the crucial feature that there is a potential linkage between the output port in of User and its input in Cop, permitting information to circulate from User to Cop when communication takes place. However, this port is still available for other users. Both users in Cop | User | User are able to communicate at different times with Cop, as illustrated in Figure 1.8

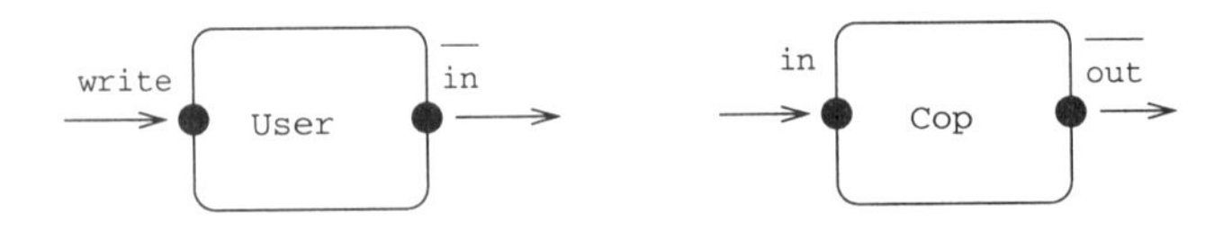

FIGURE 1.6. Flow graphs of User and Cop

[4]Equivalences between processes is discussed in Chapter 3.

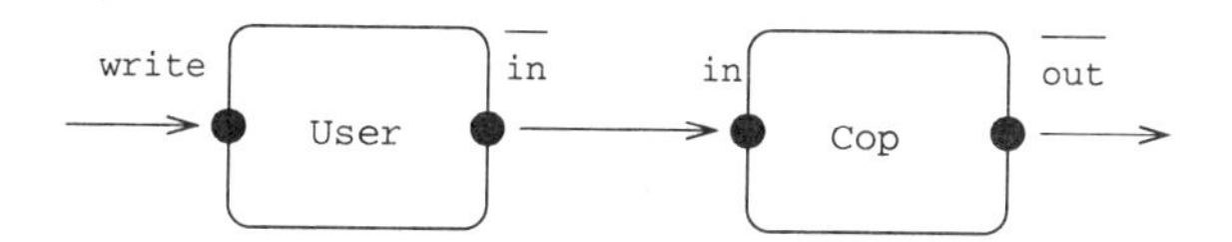

FIGURE 1.7. Flow graph of Cop | User

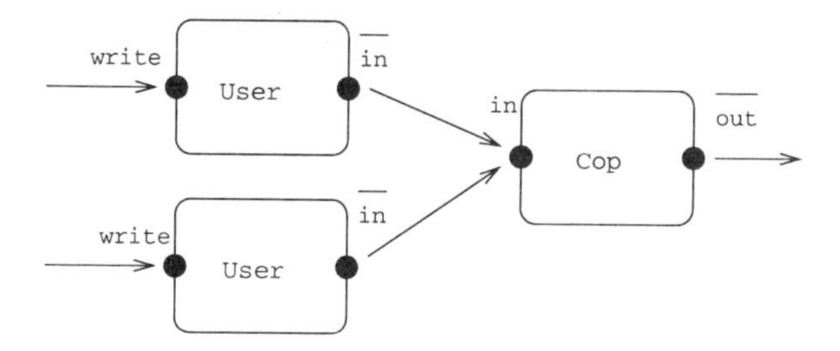

FIGURE 1.8. Flow graph of Cop | User | User

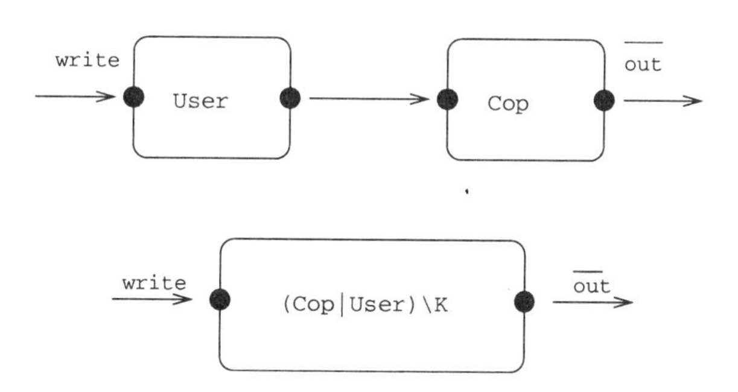

FIGURE 1.9. Flow graph of (Cop | User)$\backslash K$

The situation in which a user has private access to a copier is modelled using an abstraction or encapsulation operator that conceals ports. CCS has a *restriction* operator $\backslash J$, where J ranges over families of incomplete actions (thereby excluding the complete action τ). If K is $\{\mathtt{in}(v) \,:\, v \in D\}$ when D contains the values that can flow through in, then the port in within (Cop | User)$\backslash K$ is inaccessible to other users. The flow graph of (Cop | User)$\backslash K$ is pictured in Figure 1.9, where the linkage without names at the ports represents their concealment from other users, so it can be simplified as in the second diagram of the figure.

The visual effect of $\backslash K$ on the flow graph in Figure 1.9 is justified by the transition rule for restriction, which is as follows where $\overline{J}$ is $\{\overline{a} \,:\, a \in J\}$.

$$\mathrm{R}(\backslash)\ \ \frac{E\backslash J \stackrel{a}{\longrightarrow} F\backslash J}{E \stackrel{a}{\longrightarrow} F}\ \ a \notin J \cup \overline{J}$$

The behaviour of $E\backslash J$ is part of that of E, as any action that $E\backslash J$ may carry out can also be performed by E, but not necessarily the other way round. For instance, $\texttt{Cop} \mid \texttt{User}$ is able to perform an $\texttt{in}$ input action, whereas an attempt to derive an $\texttt{in}$ transition from $(\texttt{Cop} \mid \texttt{User})\backslash K$ is precluded because of the side condition on the rule for R($\backslash$). The presence of $\backslash K$ in $(\texttt{Cop} \mid \texttt{User})\backslash K$ prevents Cop from ever doing an in transition, except in the context of a communication with User. Restriction can therefore be used to enforce communication between parallel components. After the initial write transition $(\texttt{Cop} \mid \texttt{User})\backslash K \xrightarrow{\texttt{write}(v)} (\texttt{Cop} \mid \texttt{User}_{\texttt{v}})\backslash K$, the next transition must be a communication.

$$
\dfrac{\dfrac{\dfrac{\dfrac{}{\texttt{in}(x).\overline{\texttt{out}}(x).\texttt{Cop} \xrightarrow{\texttt{in}(v)} \overline{\texttt{out}}(v).\texttt{Cop}}}{\texttt{Cop} \xrightarrow{\texttt{in}(v)} \overline{\texttt{out}}(v).\texttt{Cop}} \quad \dfrac{\overline{\texttt{in}}(v).\texttt{User} \xrightarrow{\overline{\texttt{in}}(v)} \texttt{User}}{\texttt{User}_{\texttt{v}} \xrightarrow{\overline{\texttt{in}}(v)} \texttt{User}}}{\texttt{Cop} \mid \texttt{User}_{\texttt{v}} \xrightarrow{\tau} \overline{\texttt{out}}(v).\texttt{Cop} \mid \texttt{User}}}{(\texttt{Cop} \mid \texttt{User}_{\texttt{v}})\backslash K \xrightarrow{\tau} (\overline{\texttt{out}}(v).\texttt{Cop} \mid \texttt{User})\backslash K}
$$

A port a is concealed by restricting all the actions $\{a(v) : v \in D\}$, and therefore we shall usually abbreviate such a subset within a restriction to $\{a\}$.

Process descriptions can become quite large, especially when they consist of multiple components in parallel. We shall therefore employ abbreviations of process expressions using the relation $\equiv$, where $P \equiv F$ means that P abbreviates F, which is typically a large expression.

Example 1 The mesh of abstraction and concurrency is further revealed in the finite state example without data of a level crossing in Figure 1.10 from Bradfield and the author [10], consisting of three components Road, Rail and Signal. The actions car and train represent the approach of a car and a train, up opens the gates for the car, $\overline{\texttt{ccross}}$ is the car crossing, down closes the gates, green is the receipt of a green signal by the train, $\overline{\texttt{tcross}}$ is the train crossing, and red automatically sets the light red. Unlike most crossings, it keeps the barriers down except when a car actually approaches and tries to cross. The flow graphs of the components, and of the overall system are depicted in Figure 1.11. The transition graph is pictured in Figure 1.12. Both Road and Rail are simple cyclers that can only perform a determinate sequence of actions repeatedly.

$$
\begin{aligned}
\texttt{Road} &\stackrel{\text{def}}{=} \texttt{car.up.}\overline{\texttt{ccross}}.\overline{\texttt{down}}.\texttt{Road} \\
\texttt{Rail} &\stackrel{\text{def}}{=} \texttt{train.green.}\overline{\texttt{tcross}}.\overline{\texttt{red}}.\texttt{Rail} \\
\texttt{Signal} &\stackrel{\text{def}}{=} \overline{\texttt{green}}.\texttt{red.Signal} + \overline{\texttt{up}}.\texttt{down.Signal}
\end{aligned}
$$

$$
\texttt{Crossing} \equiv (\texttt{Road} \mid \texttt{Rail} \mid \texttt{Signal})\backslash\{\texttt{green}, \texttt{red}, \texttt{up}, \texttt{down}\}
$$

FIGURE 1.10. A level crossing

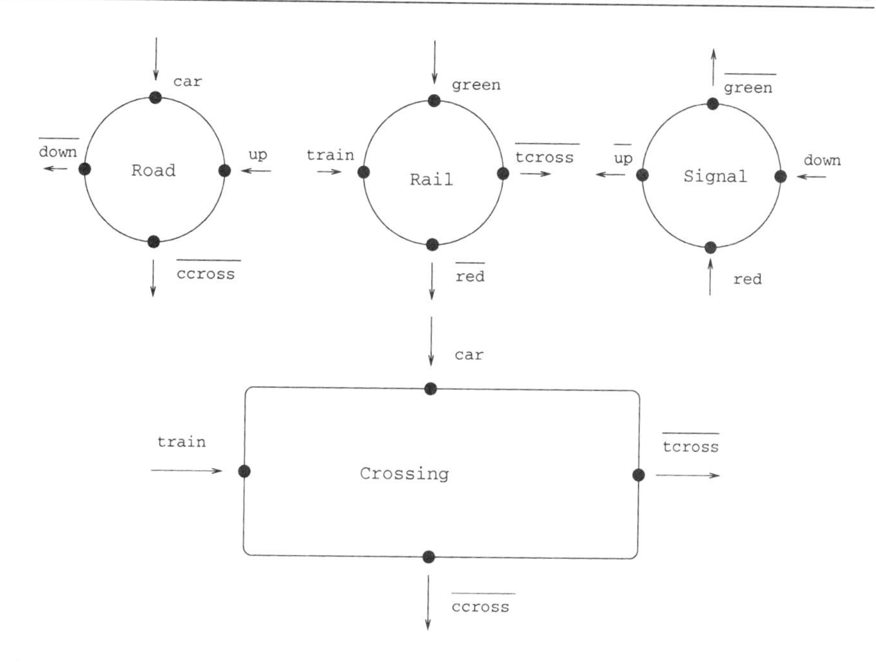

FIGURE 1.11. Flow graphs of the crossing and its components

An important arena for process descriptions is provided by modelling protocols. An example is the process Protocol of Figure 1.13 taken from Walker [60], which models an extremely simple communications protocol that allows a message to be lost during transmission. Its flow graph is the same as that of Cop, and the size of its transition graph depends on the space of messages. The sender transmits any message it receives at the port in to the medium. In turn, the medium may transmit the message to the receiver, or instead the message may be lost, an action modelled as the silent τ action, in which case the medium sends a timeout signal to the sender and the message is retransmitted. On receiving a message, the receiver transmits it at the port out and then sends an acknowledgement directly to the sender (which we assume can not be lost). Having received the acknowledgement, the sender may again receive a message at port in.

Although the flow graphs for Protocol and Cop are the same, their levels of detail are very different. The process Cop is a one-place buffer that takes in a value and later expels it. Similarly, the protocol takes in a message and later may output it. The transition graph associated with this process when there is just one message is pictured in Figure 1.14. It turns out that Protocol and Cop are observationally equivalent, as defined in Chapter 3. As process descriptions, however, they are very different. Cop is close to a specification, as its desired behaviour is given merely in terms of what it does. In contrast, Protocol is closer to an implementation, because it is defined in terms of how it is built from simpler components.

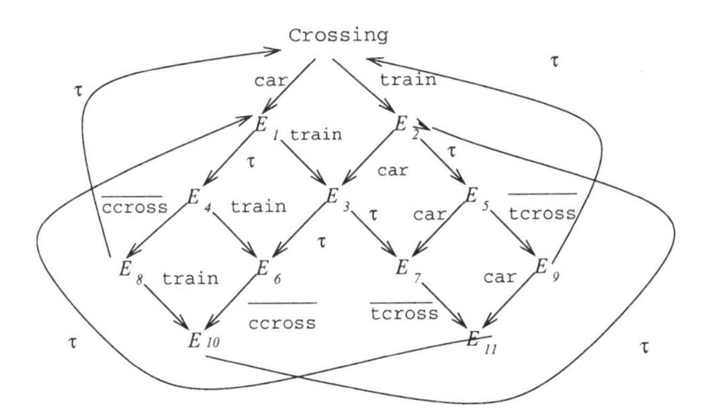

$$
\begin{aligned}
K &= \{\mathtt{green}, \mathtt{red}, \mathtt{up}, \mathtt{down}\}\\[1em]
E_1 &\equiv (\mathtt{up}.\overline{\mathtt{ccross}}.\overline{\mathtt{down}}.\mathtt{Road} \mid \mathtt{Rail} \mid \mathtt{Signal})\backslash K\\
E_2 &\equiv (\mathtt{Road} \mid \mathtt{green}.\overline{\mathtt{tcross}}.\overline{\mathtt{red}}.\mathtt{Rail} \mid \mathtt{Signal})\backslash K\\
E_3 &\equiv (\mathtt{up}.\overline{\mathtt{ccross}}.\overline{\mathtt{down}}.\mathtt{Road} \mid \mathtt{green}.\overline{\mathtt{tcross}}.\overline{\mathtt{red}}.\mathtt{Rail} \mid \mathtt{Signal})\backslash K\\
E_4 &\equiv (\overline{\mathtt{ccross}}.\overline{\mathtt{down}}.\mathtt{Road} \mid \mathtt{Rail} \mid \mathtt{down}.\mathtt{Signal})\backslash K\\
E_5 &\equiv (\mathtt{Road} \mid \overline{\mathtt{tcross}}.\overline{\mathtt{red}}.\mathtt{Rail} \mid \mathtt{red}.\mathtt{Signal})\backslash K\\
E_6 &\equiv (\overline{\mathtt{ccross}}.\overline{\mathtt{down}}.\mathtt{Road} \mid \mathtt{green}.\overline{\mathtt{tcross}}.\overline{\mathtt{red}}.\mathtt{Rail} \mid \mathtt{down}.\mathtt{Signal})\backslash K\\
E_7 &\equiv (\mathtt{up}.\overline{\mathtt{ccross}}.\overline{\mathtt{down}}.\mathtt{Road} \mid \overline{\mathtt{tcross}}.\overline{\mathtt{red}}.\mathtt{Rail} \mid \mathtt{red}.\mathtt{Signal})\backslash K\\
E_8 &\equiv (\overline{\mathtt{down}}.\mathtt{Road} \mid \mathtt{Rail} \mid \mathtt{down}.\mathtt{Signal})\backslash K\\
E_9 &\equiv (\mathtt{Road} \mid \overline{\mathtt{red}}.\mathtt{Rail} \mid \mathtt{red}.\mathtt{Signal})\backslash K\\
E_{10} &\equiv (\overline{\mathtt{down}}.\mathtt{Road} \mid \mathtt{green}.\overline{\mathtt{tcross}}.\overline{\mathtt{red}}.\mathtt{Rail} \mid \mathtt{down}.\mathtt{Signal})\backslash K\\
E_{11} &\equiv (\mathtt{up}.\overline{\mathtt{ccross}}.\overline{\mathtt{down}}.\mathtt{Road} \mid \overline{\mathtt{red}}.\mathtt{Rail} \mid \mathtt{red}.\mathtt{Signal})\backslash K
\end{aligned}
$$

FIGURE 1.12. Transition graph of Crossing

$$
\begin{aligned}
\mathtt{Sender} &\stackrel{\mathrm{def}}{=} \mathtt{in}(x).\overline{\mathtt{sm}}(x).\mathtt{Send1}(x)\\
\mathtt{Send1}(x) &\stackrel{\mathrm{def}}{=} \mathtt{ms}.\overline{\mathtt{sm}}(x).\mathtt{Send1}(x) + \mathtt{ok}.\mathtt{Sender}\\
\mathtt{Medium} &\stackrel{\mathrm{def}}{=} \mathtt{sm}(y).\mathtt{Med1}(y)\\
\mathtt{Med1}(y) &\stackrel{\mathrm{def}}{=} \overline{\mathtt{mr}}(y).\mathtt{Medium} + \tau.\overline{\mathtt{ms}}.\mathtt{Medium}\\
\mathtt{Receiver} &\stackrel{\mathrm{def}}{=} \mathtt{mr}(x).\overline{\mathtt{out}}(x).\overline{\mathtt{ok}}.\mathtt{Receiver}\\[1em]
\mathtt{Protocol} &\equiv (\mathtt{Sender} \mid \mathtt{Medium} \mid \mathtt{Receiver})\backslash\{\mathtt{sm}, \mathtt{ms}, \mathtt{mr}, \mathtt{ok}\}
\end{aligned}
$$

FIGURE 1.13. A simple protocol

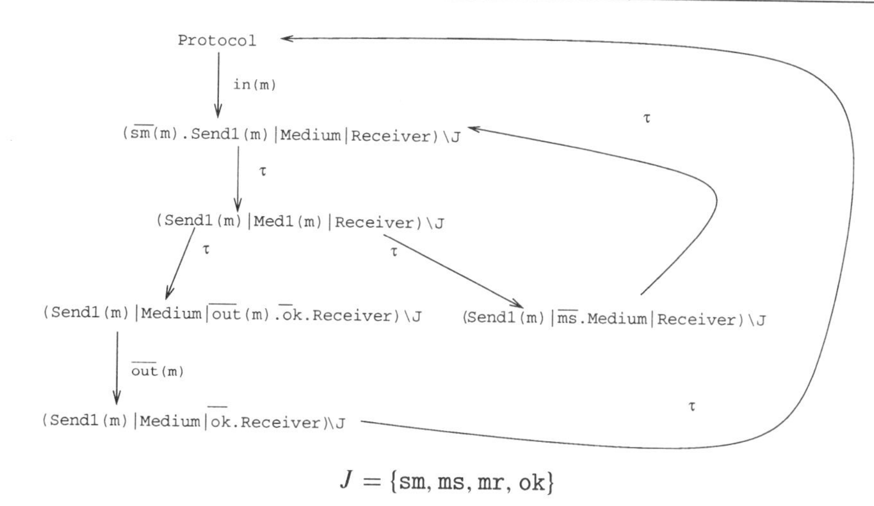

FIGURE 1.14. Protocol transition graph when there is one message m.

$$\begin{aligned}
\texttt{IO} &\stackrel{\text{def}}{=} \texttt{slot}.\overline{\texttt{bank}}.(\texttt{lost}.\overline{\texttt{loss}}.\texttt{IO} + \texttt{release}(y).\overline{\texttt{win}}(y).\texttt{IO}) \\
\texttt{B}_\texttt{n} &\stackrel{\text{def}}{=} \texttt{bank}.\overline{\texttt{max}}(n+1).\texttt{left}(y).\texttt{B}_y \\
\texttt{D} &\stackrel{\text{def}}{=} \texttt{max}(z).(\overline{\texttt{lost}}.\overline{\texttt{left}}(z).\texttt{D} + \textstyle\sum\{\overline{\texttt{release}}(y).\overline{\texttt{left}}(z-y).\texttt{D} \;:\; 1 \le y \le z\})
\end{aligned}$$

$$\texttt{SM}_\texttt{n} \equiv (\texttt{IO} \mid \texttt{B}_\texttt{n} \mid \texttt{D})\backslash\{\texttt{bank}, \texttt{lost}, \texttt{max}, \texttt{left}, \texttt{release}\}$$

FIGURE 1.15. A slot machine

Example 2 An example of an infinite state system from Bradfield and the author [10] is the slot machine $\texttt{SM}_n$ defined in Figure 1.15. Its flow graph is also depicted there. A coin is input (the action `slot`) and then, after some silent activity, either a loss or a winning sum of money is output. The system consists of three components: `IO`, which handles the taking and paying out of money; $\texttt{B}_\texttt{n}$, a bank holding n pounds; and `D`, the wheel-spinning decision component.

Exercises

1. Give a derivation of the following transition.

$$\texttt{Cop} \mid (\texttt{User}_{\texttt{v1}} \mid \texttt{User}_{\texttt{v2}}) \stackrel{\tau}{\longrightarrow} \overline{\texttt{out}}(v2).\texttt{Cop} \mid (\texttt{User}_{\texttt{v1}} \mid \texttt{User})$$

2. Show that the following three processes

   a. $(\texttt{Cop} \mid \texttt{User}_{\texttt{v1}}) \mid \texttt{User}_{\texttt{v2}}$

   b. $\texttt{User}_{\texttt{v1}} \mid (\texttt{Cop} \mid \texttt{User}_{\texttt{v2}})$

   c. $\texttt{Cop} \mid (\texttt{User}_{\texttt{v1}} \mid \texttt{User}_{\texttt{v2}})$

   have isomorphic transition graphs (and flow graphs).

3. $\texttt{Sem} \stackrel{\text{def}}{=} \texttt{get.put.Sem}$ is a semaphore. Draw the transition graph for $\texttt{Sem} \mid \texttt{Sem} \mid \texttt{Sem} \mid \texttt{Sem}$.

4. How does the transition graph for $\texttt{Cnt}$ differ from that for the counter $\texttt{Ct}_0$ of Figure 1.4?

5. Draw the transition graph for $\text{Bag} \stackrel{\text{def}}{=} \texttt{in}(x).(\overline{\texttt{out}}(x).0 \mid \text{Bag})$ when the space of values contains just two elements, 0 and 1.

6. Let L_1 be the set of actions $\{\texttt{1p}, \texttt{little}\}$ and let L_2 be $\{\texttt{1p}, \texttt{little}, \texttt{2p}\}$. Also let $\texttt{Use}_1 \stackrel{\text{def}}{=} \overline{\texttt{1p}}.\overline{\texttt{little}}.\texttt{Use}_1$. Draw flow graphs and transition graphs for the processes

   a. $\texttt{Ven} \mid \texttt{Use}_1$

   b. $\texttt{Ven} \mid (\texttt{Use}_1 \mid \texttt{Use}_1)$

   c. $(\texttt{Ven} \mid \texttt{Use}_1)\backslash L_i$

   d. $(\texttt{Ven} \mid \texttt{Use}_1)\backslash L_i \mid \texttt{Use}_1$

   e. $(\texttt{Ven} \mid \texttt{Use}_1 \mid \texttt{Use}_1)\backslash L_i$

   when $i = 1$ and $i = 2$.

7. Let $\mathcal{G}(E)$ be the transition graph for E. Define prefixing (.), $+$, $\mid$ and $\backslash J$ operators directly on transition graphs so that each of the following pairs is isomorphic.

   a. $a.\mathcal{G}(E)$ and $\mathcal{G}(a.E)$

   b. $\mathcal{G}(E + F)$ and $\mathcal{G}(E) + \mathcal{G}(F)$

   c. $\mathcal{G}(E \mid F)$ and $\mathcal{G}(E) \mid \mathcal{G}(F)$

   d. $\mathcal{G}(E)\backslash J$ and $\mathcal{G}(E\backslash J)$

8. Consider the definition of the following process from Hennessy and Ingolfsdottir [27].

$$\begin{aligned} \texttt{Fac} \stackrel{\text{def}}{=}\ & \texttt{in}_1(y).\texttt{in}_2(z).\textbf{if } y = 0 \textbf{ then } \overline{\texttt{out}}(z).0 \\ & \qquad \textbf{else } (\overline{\texttt{in}_1}(y-1).\overline{\texttt{in}_2}(z * y).0 \mid \texttt{Fac}) \end{aligned}$$

   Draw the transition graph of $(\overline{\texttt{in}_1}(3).\overline{\texttt{in}_2}(1).0 \mid \texttt{Fac})\backslash\{\texttt{in}_1, \texttt{in}_2\}$.

9. Draw the transition graph for $\texttt{Road} \mid \texttt{Rail} \mid \texttt{Signal}$, and compare it with that for $\texttt{Crossing}$.

10. Draw flow and transition graphs for the components of $\texttt{Protocol}$.

11. Refine the description of Protocol so that acknowledgements may also be lost.

1.3 Observable transitions

Actions a on the transition relations $\xrightarrow{a}$ between processes can be extended to finite length sequences w, which are also called "traces." The extended transition $E \xrightarrow{w} F$ states that E may perform the trace w and become F. There are two transition rules for traces, where ε is the empty sequence of actions.

$$\text{R(tr)} \quad E \xrightarrow{\varepsilon} E \qquad \frac{E \xrightarrow{aw} F}{E \xrightarrow{a} E' \quad E' \xrightarrow{w} F}$$

First is the axiom that any process may carry out the empty sequence and remain unchanged. The second rule allows traces to be extended. If $E \xrightarrow{a} E'$ and E' can perform the trace w and become F then $E \xrightarrow{aw} F$. No distinction is made between carrying out the action a and carrying out the trace a (understood as an action sequence of length one). Below is the derivation of the extended transition $\mathtt{Ven_b} \xrightarrow{\mathtt{big\,collect_b}} \mathtt{Ven}$ when $\mathtt{Ven_b}$ is part of the vending machine of Section 1.1.

$$\frac{\mathtt{Ven_b} \xrightarrow{\mathtt{big\,collect_b}} \mathtt{Ven}}{\dfrac{\mathtt{Ven_b} \xrightarrow{\mathtt{big}} \mathtt{collect_b.Ven}}{\mathtt{big.collect_b.Ven} \xrightarrow{\mathtt{big}} \mathtt{collect_b.Ven}} \qquad \mathtt{collect_b.Ven} \xrightarrow{\mathtt{collect_b}} \mathtt{Ven}}$$

Internal τ actions have a different status from incomplete actions. An incomplete action is "observable" because it is susceptible of interaction in a parallel context. Suppose that E may at some time perform the action ok, and that Resource is a resource. In the context $(E \mid \overline{\mathtt{ok}}.\mathtt{Resource})\backslash\{\mathtt{ok}\}$ access to Resource is only triggered with an execution of ok by E. Observation of ok is the same as the release of Resource. The silent action τ cannot be observed in this way. Consequently, an important abstraction of process behaviour derives from silent activity.

Consider the following copier C and the user U.

$$\begin{aligned} \mathtt{C} &\stackrel{\text{def}}{=} \mathtt{in}(x).\overline{\mathtt{out}}(x).\overline{\mathtt{ok}}.\mathtt{C} \\ \mathtt{U} &\stackrel{\text{def}}{=} \mathtt{write}(x).\overline{\mathtt{in}}(x).\mathtt{ok}.\mathtt{U} \end{aligned}$$

U writes a file before sending it through in and then waits for an acknowledgement. $(\mathtt{C} \mid \mathtt{U})\backslash\{\mathtt{in}, \mathtt{ok}\}$ has similar behaviour to Ucop.

$$\mathtt{Ucop} \stackrel{\text{def}}{=} \mathtt{write}(x).\overline{\mathtt{out}}(x).\mathtt{Ucop}$$

The only difference in their abilities is internal activity. Both are initially able only to carry out a write action

$$\mathtt{Ucop} \stackrel{\mathtt{write}(v)}{\longrightarrow} \overline{\mathtt{out}}(v).\mathtt{Ucop}$$

$$(\mathtt{C} \mid \mathtt{U})\backslash\{\mathtt{in}, \mathtt{ok}\} \stackrel{\mathtt{write}(v)}{\longrightarrow} (\mathtt{C} \mid \overline{\mathtt{in}}(v).\mathtt{ok}.\mathtt{U})\backslash\{\mathtt{in}, \mathtt{ok}\}.$$

Process $\overline{\mathtt{out}}(v).\mathtt{Ucop}$ outputs immediately, whereas the other process must first perform a communication before it outputs, and then τ again before a second write can happen. By abstracting from silent behaviour, this difference disappears. Outwardly, both processes repeatedly write and output.

A trace w is a sequence of actions. The trace $w \upharpoonright J$ is the subsequence of w when actions that do not belong to J are erased.

$$\varepsilon \upharpoonright J = \varepsilon$$

$$aw \upharpoonright J = \begin{cases} a(w \upharpoonright J) & \text{if } a \in J \\ w \upharpoonright J & \text{otherwise} \end{cases}$$

Below are three simple examples.

$$(\mathtt{train}\ \tau\ \overline{\mathtt{tcross}}\ \tau) \upharpoonright \{\overline{\mathtt{tcross}}\} = \overline{\mathtt{tcross}}$$

$$(\tau\ \overline{\mathtt{ccross}}\ \tau) \upharpoonright \{\overline{\mathtt{tcross}}\} = \varepsilon$$

$$(\mathtt{write}(v)\ \tau\ \overline{\mathtt{out}}(v)\ \tau) \upharpoonright \{\mathtt{write}, \overline{\mathtt{out}}\} = \mathtt{write}(v)\ \overline{\mathtt{out}}(v)$$

Associated with any trace w is the *observable* trace $w \upharpoonright \mathsf{O}$, where O is a universal set of observable actions containing at least all actions mentioned in this work apart from τ. The effect of $\upharpoonright \mathsf{O}$ on w is to erase all occurrences of the silent action τ, as illustrated by the following examples.

$$(\mathtt{in}(m)\ \tau\ \tau\ \overline{\mathtt{out}}(m)\ \tau) \upharpoonright \mathsf{O} = \mathtt{in}(m)\ \overline{\mathtt{out}}(m)$$

$$(\mathtt{in}(m)\ \tau\ \tau\ \tau\ \tau) \upharpoonright \mathsf{O} = \mathtt{in}(m)$$

$$(\tau\ \tau\ \tau\ \tau\ \tau\ \tau) \upharpoonright \mathsf{O} = \varepsilon$$

To capture observable behaviour, another family of transition relations between processes is introduced. $E \stackrel{u}{\Longrightarrow} F$ expresses that E may carry out the observable trace u and become F. The transition rule for observable traces is as follows.

$$\text{R(Tr)} \quad \frac{E \stackrel{u}{\Longrightarrow} F}{E \stackrel{w}{\longrightarrow} F} \quad u = w \upharpoonright \mathsf{O}$$

An example is $\mathtt{Protocol} \stackrel{\mathtt{in}(m)\,\overline{\mathtt{out}}(m)}{\Longrightarrow} \mathtt{Protocol}$, whose derivation utilises the extended transition $\mathtt{Protocol} \stackrel{\mathtt{in}(m)\,\tau\,\tau\,\overline{\mathtt{out}}(m)\,\tau}{\longrightarrow} \mathtt{Protocol}$.

Observable traces can also be built from their component observable actions. The extended transition $\mathtt{Crossing} \stackrel{\mathtt{train}\,\overline{\mathtt{tcross}}}{\Longrightarrow} \mathtt{Crossing}$ is the result of gluing together $\mathtt{Crossing} \stackrel{\mathtt{train}}{\Longrightarrow} E$ and $E \stackrel{\overline{\mathtt{tcross}}}{\Longrightarrow} \mathtt{Crossing}$ when the intermediate state E

is E_2 or E_5 of Figure 1.12. Observable behaviour is constructed from transitions $E \stackrel{\varepsilon}{\Longrightarrow} F$ or $E \stackrel{a}{\Longrightarrow} F$ when $a \in \mathsf{O}$, whose rules are as follows.

$$\text{R}(\stackrel{\varepsilon}{\Longrightarrow}) \quad E \stackrel{\varepsilon}{\Longrightarrow} E \qquad \frac{E \stackrel{\varepsilon}{\Longrightarrow} F}{E \stackrel{\tau}{\longrightarrow} E' \quad E' \stackrel{\varepsilon}{\Longrightarrow} F}$$

$$\text{R}(\stackrel{a}{\Longrightarrow}) \quad \frac{E \stackrel{a}{\Longrightarrow} F}{E \stackrel{\varepsilon}{\Longrightarrow} E' \quad E' \stackrel{a}{\longrightarrow} F' \quad F' \stackrel{\varepsilon}{\Longrightarrow} F}$$

$E \stackrel{\varepsilon}{\Longrightarrow} F$ if E can silently evolve to F and $E \stackrel{a}{\Longrightarrow} F$ if E can silently evolve to a process that carries out a and then silently becomes F.

Example 1 The derivation of $\texttt{Protocol} \stackrel{\texttt{in}(m)}{\Longrightarrow} F_3$, where F_3 abbreviates $(\texttt{Send1}(m) \mid \texttt{Medium} \mid \overline{\texttt{out}}(m).\overline{\texttt{ok}}.\texttt{Receiver})\backslash\{\texttt{sm}, \texttt{ms}, \texttt{mr}, \texttt{ok}\}$, uses the following two intermediate states (see Figure 1.14).

$$F_1 \equiv (\overline{\texttt{sm}}(m).\texttt{Send1}(m) \mid \texttt{Medium} \mid \texttt{Receiver})\backslash\{\texttt{sm}, \texttt{ms}, \texttt{mr}, \texttt{ok}\}$$
$$F_2 \equiv (\texttt{Send1}(m) \mid \texttt{Med1}(m) \mid \texttt{Receiver})\backslash\{\texttt{sm}, \texttt{ms}, \texttt{mr}, \texttt{ok}\}$$

Below is part of the derivation.

$$\frac{\texttt{Protocol} \stackrel{\texttt{in}(m)}{\Longrightarrow} F_3}{\texttt{Protocol} \stackrel{\varepsilon}{\Longrightarrow} \texttt{Protocol} \quad \texttt{Protocol} \stackrel{\texttt{in}(m)}{\longrightarrow} F_1 \quad F_1 \stackrel{\varepsilon}{\Longrightarrow} F_3}$$

Part of the derivation of $F_1 \stackrel{\varepsilon}{\Longrightarrow} F_3$ is as follows.

$$\frac{F_1 \stackrel{\varepsilon}{\Longrightarrow} F_3}{F_1 \stackrel{\tau}{\longrightarrow} F_2 \qquad \dfrac{F_2 \stackrel{\varepsilon}{\Longrightarrow} F_3}{F_2 \stackrel{\tau}{\longrightarrow} F_3 \quad F_3 \stackrel{\varepsilon}{\Longrightarrow} F_3}}$$

Observable behaviour of a process can also be visually encapsulated as a transition graph. As in Section 1.1, ingredients of this graph are process terms related by transitions. Each edge has the form $\stackrel{\varepsilon}{\Longrightarrow}$ or $\stackrel{a}{\Longrightarrow}$ when $a \in \mathsf{O}$. Assuming a value spacc with just one element v, the observable transition graphs for $(\texttt{C} \mid \texttt{U})\backslash\{\texttt{in}, \texttt{ok}\}$ and $\texttt{Ucop}$ are pictured in Figure 1.16 (where thick arrows are used instead of $\Longrightarrow$).

There are *two* behaviour graphs associated with any process. Although both graphs contain the same vertices, they differ in their labelled edges. Observable graphs are more complex, since they contain more transitions. However, this abundance of transitions may result in redundant vertices. Figure 1.16 exemplifies this condition in the case of $(\texttt{C} \mid \texttt{U})\backslash\{\texttt{in}, \texttt{ok}\}$. The states labelled 1 and 4 have identical capabilities, as do the states labelled 2 and 3. When minimized with respect to observable equivalences, as defined in Chapter 3, these graphs may be dramatically simplified as their vertices are fused.

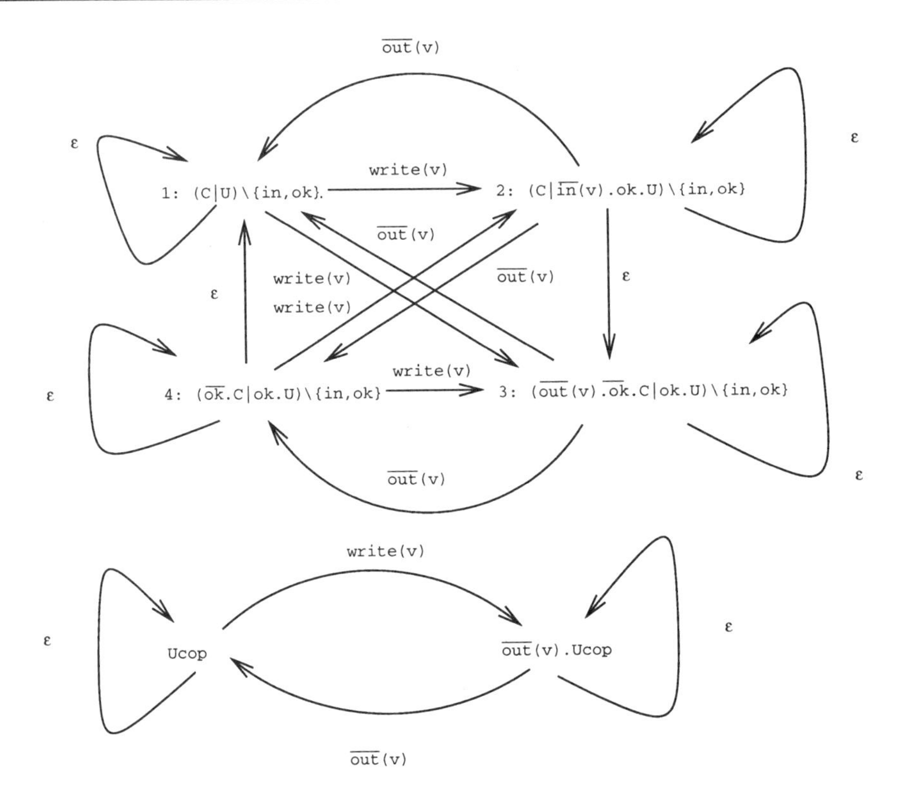

FIGURE 1.16. Observable transition graphs for (C | U)\{in, ok} and Ucop

Exercises

1. Derive the extended transition $\mathtt{SM_n} \xrightarrow{w} \mathtt{SM_{n+1}}$ when w is the following trace $\mathtt{slot}\,\tau\,\tau\,\tau\,\overline{\mathtt{loss}}$ and $\mathtt{SM_n}$ is the slot machine.
2. Provide a full derivation of $\mathtt{Protocol} \xrightarrow{s} \mathtt{Protocol}$ when s is the trace $\mathtt{in}(m)\,\tau\,\tau\,\overline{\mathtt{out}}(m)\,\tau$.
3. List the members of the following sets:

$$\{E : \mathtt{Crossing} \overset{\mathtt{train}\,\overline{\mathtt{tcross}}}{\Longrightarrow} E\}$$

$$\{E : \mathtt{Protocol} \overset{\mathtt{in}(m)}{\Longrightarrow} E\}$$

4. Show that $E \overset{a}{\Longrightarrow} F$ is derivable via the rules R(tr) and R(Tr) iff it is derivable using the rules $\mathrm{R}(\overset{a}{\Longrightarrow})$ and $\mathrm{R}(\overset{\varepsilon}{\Longrightarrow})$.
5. Draw the observable transition graphs for the processes: Cl, Ven and Crossing.
6. Although observable traces abstract from silent activity, this does not mean that internal actions can not contribute to differences in observable capability.

Let $\mathtt{Ven}'$ be a vending machine very similar to Ven of Figure 1.2, except that the initial 2p action is prefaced by the silent action, $\mathtt{Ven}' \stackrel{\text{def}}{=} \tau.\mathtt{2p.Ven_b} + \mathtt{1p.Ven_l}$

a. Show that Ven and $\mathtt{Ven}'$ have the same observable traces.

b. Let $\mathtt{Use_1}$ be the user $\mathtt{Use_1} \stackrel{\text{def}}{=} \overline{\mathtt{1p}}.\overline{\mathtt{little}}.\mathtt{Use_1}$, who is only interested in inserting the smaller coin. Show that the process $(\mathtt{Ven}' \mid \mathtt{Use_1})\backslash\{\mathtt{1p, 2p, little}\}$ may deadlock before an observable action is carried out unlike $(\mathtt{Ven} \mid \mathtt{Use_1})\backslash\{\mathtt{1p, 2p, little}\}$.

c. Draw both kinds of transition graphs for each of the processes in part (b).

7. Assuming just one datum value, draw the observable graphs for processes $(\mathtt{Cop} \mid \mathtt{User})\backslash\{\mathtt{in}\}$ and Protocol. What states of these graphs can be fused together?

8. Let $\mathcal{G}(E)$ be the transition graph for E, and let $\mathcal{G}^o(E)$ be its observable transition graph. Define the graph transformation ^{o} that maps $\mathcal{G}(E)$ into $\mathcal{G}^o(E)$.

9. A process is said to be "divergent" if it can perform the τ action forever.

a. Draw both kinds of transition graph for the following pair of processes, $\tau.0$ and $\mathtt{Div}' \stackrel{\text{def}}{=} \tau.\mathtt{Div}' + \tau.0$.

b. Do you think that the processes Protocol and Cop have the same observable behaviour? Give reasons for and against.

1.4 Renaming and linking

Cop, User and Ucop of previous sections are essentially one-place buffers, taking in a value and later expelling it. Assume that B is the following canonical buffer.

$$\mathtt{B} \stackrel{\text{def}}{=} \mathtt{i}(x).\overline{\mathtt{o}}(x).\mathtt{B}$$

For instance, Cop is the process B when port i is in and port o is out. Relabelling of ports can be made explicit by introducing an operator which renames actions.

The crux of renaming is a function mapping actions into actions. To ensure pleasant properties, a renaming function f is subject to a few restrictions. First, it should respect complements. For any observable a, the actions $f(a)$ and $f(\overline{a})$ are co-actions, that is $f(\overline{a}) = \overline{f(a)}$. Second, it should conserve the silent action, $f(\tau) = \tau$. Associated with any function f obeying these conditions is the renaming operator $[f]$, which, when applied to process E, is written as $E[f]$; this is the process E whose actions are relabelled according to f.

A renaming function f can be abbreviated to its essential part. If each a_i is a distinct observable action, then $b_1/a_1, \ldots, b_n/a_n$ represents the function f that renames a_i to b_i (and $\overline{a_i}$ to $\overline{b_i}$), and leaves any other action c unchanged. For instance, Cop abbreviates the process $\mathtt{B[in/i, out/o]}$: here we maintain the convention that in stands for the family $\{\mathtt{in}(v) : v \in D\}$ and i for $\{\mathtt{i}(v) : v \in D\}$, so $\mathtt{in/i}$ symbolises the function that also preserves values by mapping $\mathtt{i}(v)$ to $\mathtt{in}(v)$ for each v. The transition rule for renaming is set forth below.

$$\text{R}([f]) \quad \frac{E[f] \xrightarrow{a} F[f]}{E \xrightarrow{b} F} \; a = f(b)$$

This rule is used in derivations of the following pair of transitions.

$$\mathtt{B[in/i, out/o]} \xrightarrow{\mathtt{in}(v)} (\overline{\mathtt{o}}(v).\mathtt{B})\mathtt{[in/i, out/o]} \xrightarrow{\overline{\mathtt{out}}(v)} \mathtt{B[in/i, out/o]}$$

Below is the derivation of the initial transition.

$$\cfrac{\mathtt{B[in/i, out/o]} \xrightarrow{\mathtt{in}(v)} (\overline{\mathtt{o}}(v).\mathtt{B})\mathtt{[in/i, out/o]}}{\cfrac{\mathtt{B} \xrightarrow{\mathtt{i}(v)} \overline{\mathtt{o}}(v).\mathtt{B}}{\mathtt{i}(x).\overline{\mathtt{o}}(x).\mathtt{B} \xrightarrow{\mathtt{i}(v)} \overline{\mathtt{o}}(v).\mathtt{B}}}$$

A virtue of process modelling is that it allows building systems from simpler components. Consider how to model an n-place buffer when $n > 1$, following Milner [44], by linking together n instances of B in parallel. The flow graph of n copies of B is pictured in Figure 1.17. For this to become an n-place buffer we need to "link," and then internalise, the contiguous $\overline{\mathtt{o}}$ and $\mathtt{i}$ ports. Renaming permits linking, as the following variants of B show.

$$\begin{array}{lll} \mathtt{B}_1 & \equiv & \mathtt{B[o_1/o]} \\ \mathtt{B}_{j+1} & \equiv & \mathtt{B[o}_j\mathtt{/i, o}_{j+1}\mathtt{/o]} \quad 1 \le j < n-1 \\ \mathtt{B}_n & \equiv & \mathtt{B[o}_{n-1}\mathtt{/i]} \end{array}$$

The flow graph of $\mathtt{B}_1 \mid \ldots \mid \mathtt{B}_n$ is also shown in Figure 1.17, and contains the intended links. The n-place buffer is the result of internalizing these contiguous links, $(\mathtt{B}_1 \mid \ldots \mid \mathtt{B}_n)\backslash\{\mathtt{o}_1, \ldots, \mathtt{o}_{n-1}\}$.

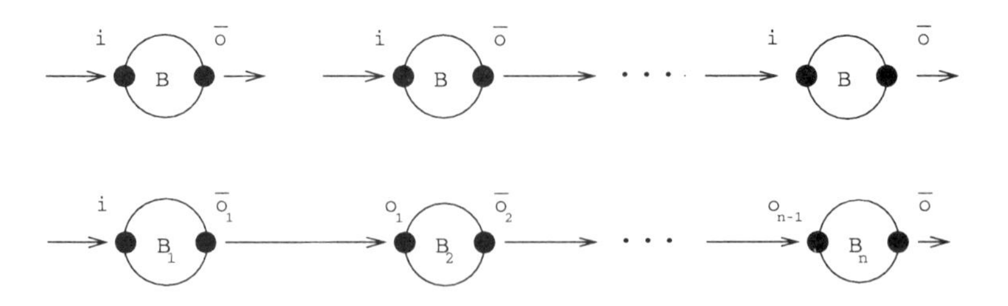

FIGURE 1.17. Flow graph of n instances of B, and $\mathtt{B}_1 \mid \ldots \mid \mathtt{B}_n$.

Part of the behaviour of a two-place buffer is illustrated by the following cycle.

$$
\begin{array}{rcl}
(\mathtt{B}[\mathtt{o}_1/\mathtt{o}] \mid \mathtt{B}[\mathtt{o}_1/\mathtt{i}])\backslash\{\mathtt{o}_1\} & \stackrel{\mathtt{i}(v)}{\longrightarrow} & ((\overline{\mathtt{o}}(v).\mathtt{B})[\mathtt{o}_1/\mathtt{o}] \mid \mathtt{B}[\mathtt{o}_1/\mathtt{i}])\backslash\{\mathtt{o}_1\} \\
 & & \downarrow \tau \\
((\overline{\mathtt{o}}(w).\mathtt{B})[\mathtt{o}_1/\mathtt{o}] \mid (\overline{\mathtt{o}}(v).\mathtt{B})[\mathtt{o}_1/\mathtt{i}])\backslash\{\mathtt{o}_1\} & \stackrel{\mathtt{i}(w)}{\longleftarrow} & (\mathtt{B}[\mathtt{o}_1/\mathtt{o}] \mid (\overline{\mathtt{o}}(v).\mathtt{B})[\mathtt{o}_1/\mathtt{i}])\backslash\{\mathtt{o}_1\} \\
\downarrow \overline{\mathtt{o}}(v) & & \\
((\overline{\mathtt{o}}(w).\mathtt{B})[\mathtt{o}_1/\mathtt{o}] \mid \mathtt{B}[\mathtt{o}_1/\mathtt{i}])\backslash\{\mathtt{o}_1\} & \stackrel{\tau}{\longrightarrow} & (\mathtt{B}[\mathtt{o}_1/\mathtt{o}] \mid (\overline{\mathtt{o}}(w).\mathtt{B})[\mathtt{o}_1/\mathtt{i}])\backslash\{\mathtt{o}_1\} \\
 & & \downarrow \overline{\mathtt{o}}(w) \\
 & & (\mathtt{B}[\mathtt{o}_1/\mathtt{o}] \mid \mathtt{B}[\mathtt{o}_1/\mathtt{i}])\backslash\{\mathtt{o}_1\}
\end{array}
$$

Below is the derivation of the second transition.

$$
\dfrac{\dfrac{\dfrac{\dfrac{}{\overline{\mathtt{o}}(v).\mathtt{B} \stackrel{\overline{\mathtt{o}}(v)}{\longrightarrow} \mathtt{B}}}{(\overline{\mathtt{o}}(v).\mathtt{B})[\mathtt{o}_1/\mathtt{o}] \stackrel{\overline{\mathtt{o}}_1(v)}{\longrightarrow} \mathtt{B}[\mathtt{o}_1/\mathtt{o}]} \quad \dfrac{\dfrac{\mathtt{i}(x).\overline{\mathtt{o}}(x).\mathtt{B} \stackrel{\mathtt{i}(v)}{\longrightarrow} \overline{\mathtt{o}}(v).\mathtt{B}}{\mathtt{B} \stackrel{\mathtt{i}(v)}{\longrightarrow} \overline{\mathtt{o}}(v).\mathtt{B}}}{\mathtt{B}[\mathtt{o}_1/\mathtt{i}] \stackrel{\mathtt{o}_1(v)}{\longrightarrow} (\overline{\mathtt{o}}(v).\mathtt{B})[\mathtt{o}_1/\mathtt{i}]}}{(\overline{\mathtt{o}}(v).\mathtt{B})[\mathtt{o}_1/\mathtt{o}] \mid \mathtt{B}[\mathtt{o}_1/\mathtt{i}] \stackrel{\tau}{\longrightarrow} \mathtt{B}[\mathtt{o}_1/\mathtt{o}] \mid (\overline{\mathtt{o}}(v).\mathtt{B})[\mathtt{o}_1/\mathtt{i}]}}{((\overline{\mathtt{o}}(v).\mathtt{B})[\mathtt{o}_1/\mathtt{o}] \mid \mathtt{B}[\mathtt{o}_1/\mathtt{i}])\backslash\{\mathtt{o}_1\} \stackrel{\tau}{\longrightarrow} (\mathtt{B}[\mathtt{o}_1/\mathtt{o}] \mid (\overline{\mathtt{o}}(v).\mathtt{B})[\mathtt{o}_1/\mathtt{i}])\backslash\{\mathtt{o}_1\}}
$$

A more involved example from Milner [44] refers to the construction of a scheduler from small cycling components. Assume n tasks when $n > 1$, and that action a_i initiates the ith task, whereas b_i signals its completion. The scheduler plans the order of task initiation, ensuring that the sequence of actions $a_1 \dots a_n$ is carried out cyclically starting with a_1. The tasks may terminate in any order, but a task can not be restarted until its previous operation has finished. So, the scheduler must guarantee that the actions a_i and b_i happen alternately for each i.

Let $\mathtt{Cy}'$ be a cycler of length four, $\mathtt{Cy}' \stackrel{\text{def}}{=} a.c.b.d.\mathtt{Cy}'$, whose flow graph is illustrated in Figure 1.18. In this case, the flow graph is very close to its transition graph, so we have circled the a label to indicate that it is initially active. As soon as a happens, control passes to the active action c. The clockwise movement of activity

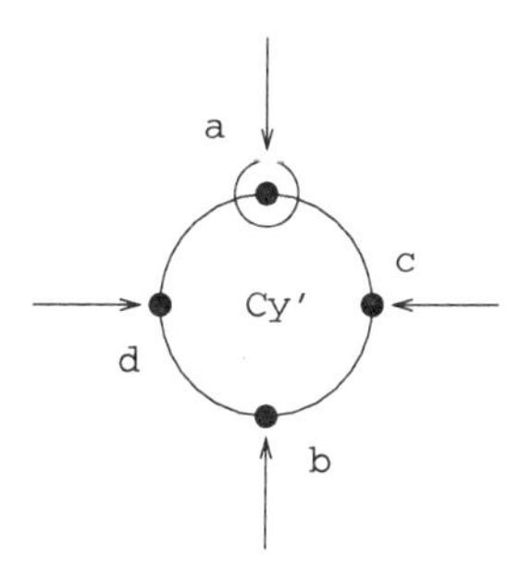

FIGURE 1.18. The flow graph of $\mathtt{Cy}'$

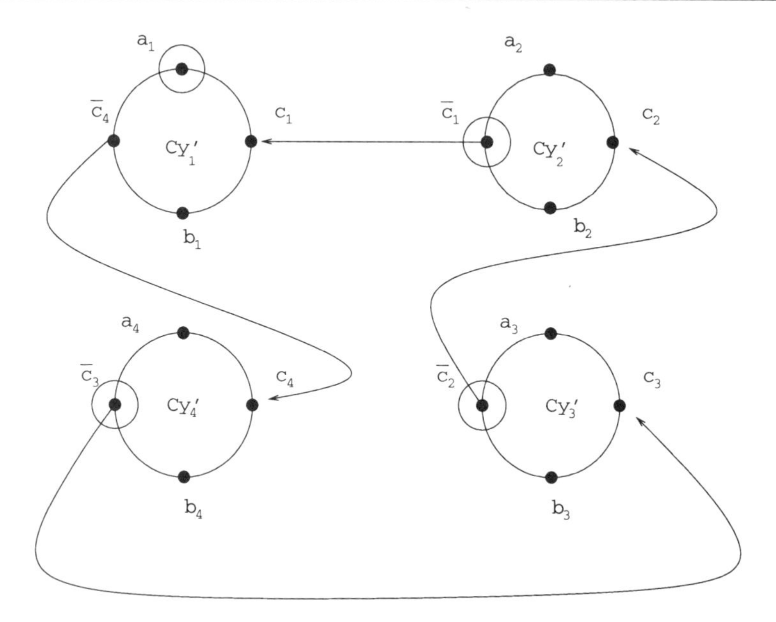

FIGURE 1.19. Flow graph of $\texttt{Cy}'_1 \mid \texttt{Cy}'_2 \mid \texttt{Cy}'_3 \mid \texttt{Cy}'_4$

around this flowgraph is its transition graph. A first attempt at building the required scheduler is as a ring of n cyclers, where the a action is task initiation, the b action is task termination, and the other actions c and d are used for synchronization.

$$\begin{array}{lcl} \texttt{Cy}'_1 & \equiv & \texttt{Cy}'[a_1/a, c_1/c, b_1/b, \overline{c}_n/d] \\ \texttt{Cy}'_i & \equiv & (d.\texttt{Cy}')[a_i/a, c_i/c, b_i/b, \overline{c}_{i-1}/d] \;\; 1 < i \leq n \end{array}$$

$\texttt{Cy}'_1$ carries out the cycle $\texttt{Cy}'_1 \stackrel{a_1\,c_1\,b_1\,\overline{c}_n}{\longrightarrow} \texttt{Cy}'_1$ and $\texttt{Cy}'_i$, for $i > 1$ carries out the different cycle $\texttt{Cy}'_i \stackrel{\overline{c}_{i-1}\,a_i\,c_i\,b_i}{\longrightarrow} \texttt{Cy}'_i$.

The flow graph of the process $\texttt{Cy}'_1 \mid \texttt{Cy}'_2 \mid \texttt{Cy}'_3 \mid \texttt{Cy}'_4$ with initial active transitions marked is pictured in Figure 1.19. Next, the c_i actions are internalised. Assume that $\texttt{Sched}'_4 \equiv (\texttt{Cy}'_1 \mid \texttt{Cy}'_2 \mid \texttt{Cy}'_3 \mid \texttt{Cy}'_4)\backslash\{c_1, \ldots, c_4\}$. Imagine that the c_i actions are concealed in Figure 1.19, and notice then how the tasks must be initiated cyclically. For example, a_3 can only happen once a_1, and then a_2, have both happened. Moreover, no task can be reinitiated until its previous execution has terminated. For example, a_3 can not recur until b_3 has happened. However, $\texttt{Sched}'_4$ does not permit all possible acceptable behaviour. Put simply, action b_4 cannot happen before b_1 because of the synchronization between c_4 and $\overline{c}_4$, meaning task four cannot terminate before the initial task.

Milner's solution in [44] to this problem is to redefine the cycler

$$\texttt{Cy} \stackrel{\text{def}}{=} a.c.(b.d.\texttt{Cy} + d.b.\texttt{Cy})$$

and to use the same renaming functions. Let $\mathtt{Cy}_i$ for $1 < i \leq n$ be the process

$$(d.\mathtt{Cy})[a_i/a, c_i/c, b_i/b, \overline{c}_{i-1}/d]$$

and let $\mathtt{Cy}_1$ be $\mathtt{Cy}[a_1/a, c_1/c, b_1/b, \overline{c}_n/d]$. The required scheduler is $\mathtt{Sched}_n$, the process $(\mathtt{Cy}_1 \mid \ldots \mid \mathtt{Cy}_n)\backslash\{c_1, \ldots, c_n\}$.

Exercises

1. Redefine Road and Rail from Section 1.2 as abbreviations of $\mathtt{Cy}'$ plus renaming.
2. Assuming that the space of values consists of one element, draw both kinds of transition graph for the three-place buffer

   $$(\mathtt{B}_1 \mid \mathtt{B}_2 \mid \mathtt{B}_3)\backslash\{\mathtt{o}_1, \mathtt{o}_2\}.$$

3. What extra condition on a renaming function f is necessary to ensure that the transition graphs of $(E \mid F)[f]$ and $E[f] \mid F[f]$ be isomorphic? Do either of the buffer and scheduler examples fulfil this condition?
4. **a.** Draw both kinds of transition graph for the processes $\mathtt{Sched}_4$ and $\mathtt{Sched}'_4$.

   b. Prove that $\mathtt{Sched}'_4$ permits all, and only the acceptable, behaviour of a scheduler (as described earlier).
5. From Milner [44]. Construct a sorting machine from simple components for each $n \geq 1$ capable of sorting n-length sequences of natural numbers greater than 0. It accepts exactly n numbers, one by one at in, then delivers them up one by one in descending order at $\overline{\mathtt{out}}$, terminated by a 0. Thereafter, it returns to its initial state.

1.5 More combinations of processes

In previous sections we have emphasised the process combinators of CCS. There is a variety of process calculi dedicated to precise modelling of systems. Besides CCS and CSP, there is ACP, due to Bergstra and Klop [5, 3], Hennessy's EPL [26], MEIJE defined by Austry, Boudol and Simone [2, 51], Milner's SCCS [43], and Winskel's general process algebra [62]. Although the behavioural meaning of all the operators of these calculi can be presented using inference rules, their conception reflects different concerns. ACP is primarily algebraic, highlighting equations[5]. CSP was devised with a distinguished model in mind, the failures model[6], and MEIJE was introduced as a very expressive calculus, initiating general results about families of transition rules that can be used to define process operators; see Groote and Vaandrager [25]. The general process algebra in [62] has roots

[5] See Section 3.6.

[6] See Section 2.2 for the notion of failure.

in category theory. Moreover, users of process notation can introduce their own operators according to the application at hand.

Numerous parallel operators are proposed within the calculi mentioned above. Their transition rules are of two kinds. First, where $\times$ is parallel, is a synchronization rule.

$$\frac{E \times F \stackrel{ab}{\longrightarrow} E' \times F'}{E \stackrel{a}{\longrightarrow} E' \quad F \stackrel{b}{\longrightarrow} F'} \ \ldots$$

Here, ab is the concurrent product of the component actions a and b, and $\ldots$ may be filled in with a side condition. In the case of the parallel of Section 1.2, the actions a and b must be co-actions, and their concurrent product is the silent action. Other rules permit components to act alone.

$$\frac{E \times F \stackrel{a}{\longrightarrow} E' \times F}{E \stackrel{a}{\longrightarrow} E'} \ \ldots \qquad \frac{E \times F \stackrel{a}{\longrightarrow} E \times F'}{F \stackrel{a}{\longrightarrow} F'} \ \ldots$$

In the case of the parallel $|$ there are no side conditions when applying these rules. This general format covers a variety of parallel operators. At one extreme is the case when $\times$ is a synchronous parallel (as in SCCS), when only the synchronization rule applies, thereby forcing maximal concurrent interaction. At the other extreme is a pure interleaving operator when the synchronization rule never applies. In between are the parallel operators of ACP, CCS and CSP.

A different conception of synchronization underlies the parallel operator of CSP (when data is not passed). Synchronization is "sharing" the same action. Actions now do not have partner co-actions because multiple parallel processes may synchronize. Each process instance in CSP has an associated alphabet consisting of the actions that it is willing to engage in. Two processes must synchronize on common actions belonging to both component alphabets. An alternative presentation, which does not require alphabets, consists of introducing a family of binary parallel operators $\|_K$ indexed by a set K of actions that have to be shared. Rules for $\|_K$ are as follows.

$$\frac{E\|_K F \stackrel{a}{\longrightarrow} E'\|_K F'}{E \stackrel{a}{\longrightarrow} E' \quad F \stackrel{a}{\longrightarrow} F'} \ a \in K$$

$$\frac{E\|_K F \stackrel{a}{\longrightarrow} E'\|_K F}{E \stackrel{a}{\longrightarrow} E'} \ a \notin K \qquad \frac{E\|_K F \stackrel{a}{\longrightarrow} E\|_K F'}{F \stackrel{a}{\longrightarrow} F'} \ a \notin K$$

The first rule requires that both components of $E\|_K F$ must share any action in K. The other pair allows components to proceed independently, so long as they perform actions outside of K.

Example 1 Assume that `Ven` is the vending machine of Section 1.1, and let `Use` $\stackrel{\text{def}}{=}$ `1p.little.collect`$_1$`.Use` be a user. The transition graph for `Ven`$\|_K$`Use` when K is the set {`1p`, `little`, `collect`$_1$} is isomorphic to that of `Ven`. The following initial transitions are allowed.

$$\texttt{Ven}\|_K\texttt{Use} \xrightarrow{\texttt{1p}} \texttt{Ven}_\texttt{l}\|_K\texttt{little.collect}_\texttt{l}\texttt{.Use}$$

$$\texttt{Ven}\|_K\texttt{Use} \xrightarrow{\texttt{2p}} \texttt{Ven}_\texttt{b}\|_K\texttt{Use}$$

Adding another user does not change the possible behaviour. The process $\texttt{Ven}\|_K\texttt{Use}\|_K\texttt{Use}$ also has an isomorphic transition graph to that of $\texttt{Ven}\|_K\texttt{Use}$, as all components must synchronize on K actions. If instead K is the set $\{\texttt{1p}, \texttt{little}, \texttt{collect}_\texttt{l}, \texttt{2p}\}$, then the graph of $\texttt{Ven}\|_K\texttt{Use}$ is isomorphic to $\texttt{Use}$, as the initial 2p transition is blocked.

The operator $\|_K$ enforces synchronization of actions in K. In CCS, all synchronization is silent. In CSP, silent activity is achieved using an abstraction or hiding operator, which we represent as $\backslash\backslash K$, and whose transition rules are as follows.

$$\frac{E\backslash\backslash K \xrightarrow{a} F\backslash\backslash K}{E \xrightarrow{a} F}\ a \notin K \qquad \frac{E\backslash\backslash K \xrightarrow{\tau} F\backslash\backslash K}{E \xrightarrow{a} F}\ a \in K$$

Hiding is also useful for abstracting from observable behaviour of processes that do not contain the sharing parallel operator.

Example 2 The scheduler of the previous section has to ensure that the sequence of actions $a_1 \ldots a_n$ happens cyclically. The observable behaviour of $\texttt{Sched}_n\backslash\backslash\{b_1, \ldots, b_n\}$ and of $\texttt{Sched}'_n\backslash\backslash\{b_1, \ldots, b_n\}$ is the infinite repetition of the sequence $a_1 \ldots a_n$. For example $\texttt{Sched}'_4\backslash\backslash\{b_1, \ldots, b_4\}$ carries out the following cycle.

$$\texttt{Sched}'_4\backslash\backslash\{b_1, \ldots, b_4\} \stackrel{a_1 a_2 a_3 a_4}{\Longrightarrow} \texttt{Sched}'_4\backslash\backslash\{b_1, \ldots, b_4\}$$

In CCS, values can be passed between ports. A more general idea is to allow ports themselves to be passed between processes. For example, the process $\texttt{in}(x).\overline{x}.E$ receives a port at in, which it may then use to synchronize on. This kind of general mechanism permits process mobility, since links may be dynamically altered as a system evolves. This facility is basic to the π-calculus, as developed by Milner, Parrow and Walker [45].

There is a variety of extensions to basic process calculi for modelling real-time phenomena, such as timeouts expressed using either action duration or delay intervals between actions, priorities among actions or among processes, spatially distributed systems using locations, and the description of stochastic behaviour using probabilistic, instead of nondeterministic, choice. Some of these extensions are useful for modelling hybrid systems that involve a mixture of discrete and continuous, and can be found in control systems for manufacturing (such as chemical plants).

Exercises

1. In ACP, sequential composition of processes $E; F$ is a primitive operator. The idea is that the behaviour of $E; F$ is that of E followed by that of F. Define transition rules for sequential composition. To what extent can sequential composition be simulated within CCS using parallel composition?
2. Draw both kinds of transition graph for the following processes.
   a. $\mathtt{Sched}'_4 \backslash\backslash \{b_1, \ldots, b_4\}$
   b. $\mathtt{Sched}_4 \backslash\backslash \{b_1, \ldots, b_4\}$
   c. $\mathtt{Sched}'_4 \backslash\backslash \{a_1, \ldots, a_4\}$
   d. $\mathtt{Sched}_4 \backslash\backslash \{a_1, \ldots, a_4\}$
3. Show how the operator $\backslash\backslash K$ can be defined in CCS.
4. A process E is determinate provided that, for any trace w if $E \stackrel{w}{\longrightarrow} E_1$ and $E \stackrel{w}{\longrightarrow} E_2$, then $E_1 = E_2$. Assume that E and F are determinate, and that K is the set of actions common to (that is, occurring in) both E and F. Show that $E \|_K F$ is also determinate, but that $E | F$ need not be.

1.6 Sets of processes

Processes can also be used to capture foundational models of computation, such as Turing machines, counter machines and parallel random-access machines. This remains true for the following restricted process language, where P ranges over process names, a over actions, and I over finite sets of indices.

$$E ::= P \mid \sum\{a_i.E_i \ : \ i \in I\} \mid E_1 \mid E_2 \mid E\backslash\{a\}$$

A process expression is either a name, a finite sum of process expressions, a parallel composition of process expressions, or a restricted process expression. A (closed) process is given as a finite family $\{P_i \stackrel{\text{def}}{=} E_i \ : \ 1 \leq i \leq n\}$ of definitions, where all the process names in each E_i belong to the set $\{P_1, \ldots, P_n\}$. For instance, see the example Count, below. Although process expressions such as the counter $\mathtt{Ct}_0$ (Figure 1.4) the register $\mathtt{Reg}_0$ (Section 1.1) and the slot machine $\mathtt{SM}_0$ (Figure 1.15) are excluded because their definitions appeal to value passing or infinite sets of indices, their observable behaviour can be "simulated" by processes belonging to this restricted process language.

As an example, consider the following finite reformulation of the counter $\mathtt{Ct}_0$, due to Taubner [57].

$$\begin{array}{lcl} \mathtt{Count} & \stackrel{\text{def}}{=} & \mathtt{round}.\mathtt{Count} + \mathtt{up}.(\mathtt{Count}_1 \mid a.\mathtt{Count})\backslash\{a\} \\ \mathtt{Count}_1 & \stackrel{\text{def}}{=} & \mathtt{down}.\overline{a}.0 + \mathtt{up}.(\mathtt{Count}_2 \mid b.\mathtt{Count}_1)\backslash\{b\} \\ \mathtt{Count}_2 & \stackrel{\text{def}}{=} & \mathtt{down}.\overline{b}.0 + \mathtt{up}.(\mathtt{Count}_1 \mid a.\mathtt{Count}_2)\backslash\{a\} \end{array}$$

The reader is invited to draw the observable transition graph for `Count` and compare it with Figure 1.4.

In the remaining chapters, we shall abstract from the behaviour of processes. However, in some cases this requires us to define families of processes that encapsulate the behaviour of some initial processes. This naturally leads to sets of processes that are "transition closed." A set of processes E is transition closed if, for any process E in E, and for any action a, and for any transition $E \stackrel{a}{\longrightarrow} F$, then F also belongs to E. For instance, the set of processes appearing in a transition graph is transition closed. In later chapters we use P to range over non-empty transition closed sets, and we use P(E) to range over transition closed sets that contain E.

There is a smallest transition closed set containing E, given by the set $\{F : E \stackrel{w}{\longrightarrow} F$ for some trace $w\}$. However, it may be computationally difficult, and in some cases undecidable, to determine this set. Instead, we can define larger transition closed sets containing E inductively on the structure of E. The resultant set is only an "estimate" of a smallest transition closed set. Consider the following definition of the set of subprocesses Sub(E) of an initial CCS process E that does not involve value passing or parameterisation.

$$
\begin{array}{lcl}
\text{Sub}(a.E) & = & \{a.E\} \cup \text{Sub}(E) \\
\text{Sub}(E + F) & = & \{E + F\} \cup \text{Sub}(E) \cup \text{Sub}(F) \\
\text{Sub}(E \mid F) & = & \{E \mid F\} \cup \{E' \mid F' : E' \in \text{Sub}(E) \text{ and } F' \in \text{Sub}(F)\} \\
\text{Sub}(E\backslash K) & = & \{E'\backslash K : E' \in \text{Sub}(E)\} \\
\text{Sub}(E[f]) & = & \{E'[f] : E' \in \text{Sub}(E)\} \\
\text{Sub}(P) & = & \{P\} \cup \text{Sub}(E) \text{ if } P \stackrel{\text{def}}{=} E
\end{array}
$$

Example 1 The set Sub(`Crossing`), where `Crossing` is defined in Figure 1.10, contains 125 elements including the following, where K is the set {`green`, `red`, `up`, `down`}.

$(\texttt{Road} \mid \texttt{Rail} \mid \overline{\texttt{up}}.\texttt{down.Signal})\backslash K$

$(\texttt{Road} \mid \overline{\texttt{tcross}}.\overline{\texttt{red}}.\texttt{Rail} \mid \texttt{Signal})\backslash K$

$(\overline{\texttt{down}}.\texttt{Road} \mid \texttt{Rail} \mid \texttt{red.Signal})\backslash K$

Only 12 of these elements belong to the smallest transition closed set containing `Crossing`; see Figure 1.12.

The above definition of Sub(E) is transition closed (as the reader can check). As an estimate of a smallest transition closed set, it may be very generous as illustrated in example 1. A more refined definition is possible which, for instance, would have the consequence that Sub($(a.E)\backslash\{a\}$) always has size one. The definition of Sub can also be extended to processes defined using parameters, or to processes containing value passing. One method is to instantiate the parameters immediately. For example, the following would capture the case for the input prefix.

$$\text{Sub}(a(x).E) \;=\; \{a(x).E\} \cup \bigcup \{\text{Sub}(E\{v/x\}) : v \in D\}$$

Another method is to first define the set of process "shapes," processes with free parameters, then to define instances of these shapes using substitution. The definition of the input prefix would now be as follows, provided that x does not also occur bound within E. We leave details to the reader.

$$\text{Sub}(a(x).E) \;=\; \{a(x).E\} \cup \text{Sub(E)}$$

Exercises

1. Draw the observable transition graph for `Count`. How does it compare with the graph for $\mathtt{Ct}_0$?
2. From Taubner [57]. Using two copies of `Count`, show how a two-counter machine can be modelled within the restricted process language of this section.
3. A process E can carry out the "completed observable trace" w if $E \stackrel{w}{\Longrightarrow} F$ and F is deadlocked, or has terminated. Assume that $\text{CT}(E)$ is the set of completed observable traces that E can carry out.
   a. Prove that, for any process E defined in the restricted process language of this section, the set $\text{CT}(E)$ is recursively enumerable.
   b. Let L be a recursively enumerable language over a finite set of observable actions (which, therefore, excludes τ). Prove that there is a process E of the restricted process language with the property $\text{CT}(E) = L$.
4. Answer the following open-ended questions.
   a. What criteria should be used for assessing the expressive power of a process language?
   b. Should there be a "canonical" process calculus?
   c. Is there a concurrent version of the Church-Turing thesis for sequential programs?
5. Consider any process E without parameters or value passing.
   a. Show that Sub(E) as defined above is transition closed.
   b. List all members of Sub(`Crossing`) and compare that listing with the smallest transition closed set P(`Crossing`).
   c. Refine the definition of Sub(E). What size does Sub(`Crossing`) have with your refined definition?
6. Extend the definition of Sub to processes containing parameters and value passing in both ways suggested in the text.

2

Modalities and Capabilities

Various examples of processes have been presented so far from a simple clock to a scheduler. In each case, a process is an expression constructed from a few process operators. Behaviour is determined by the transition rules for process combinators. These rules may involve side conditions relying on extra information. For instance, when data are involved, a partial evaluation function is used. Consequently, the ingredients of a process description are combinators, predicates and transition rules that allow us to deduce behaviour.

In this chapter, some abstractions from the overall behaviour of a process are considered. Already we have contrasted finite state from infinite state processes. The size of a process is determined by its transition graph, although under some circumstances an estimate is provided instead using Sub. Also, observable transitions marked by the thicker transition arrows $\stackrel{a}{\Longrightarrow}$ have been distinguished from their thinner counterparts $\stackrel{a}{\longrightarrow}$. We examine simple properties of processes as given by modal logics, whose formulas express process capabilities and necessities, and which can be used to focus on part of the behaviour of a process.

2.1 Hennessy-Milner logic I

A modal logic M is introduced for describing local capabilities of processes. Formulas of M are built from boolean connectives and the modal operators $[K]$ ("box K") and $\langle K \rangle$ ("diamond K") for any set of actions K. The following abstract syntax definition specifies formulas of M.

$$\Phi ::= \mathtt{tt} \mid \mathtt{ff} \mid \Phi_1 \wedge \Phi_2 \mid \Phi_1 \vee \Phi_2 \mid [K]\Phi \mid \langle K \rangle \Phi$$

A formula can be the constant true formula tt, the constant false formula ff, a conjunction of formulas $\Phi_1 \wedge \Phi_2$, a disjunction of formulas $\Phi_1 \vee \Phi_2$ or a formula $[K]\Phi$ or $\langle K \rangle \Phi$ prefaced with a modal operator.

Formulas of M are ascribed to processes. Each process either has a modal property, or fails to have it. When process E has property Φ, we write $E \models \Phi$ and when it fails to have Φ, we write $E \not\models \Phi$. If E has Φ we often say "E satisfies Φ" or "E realises Φ." The binary satisfaction relation between processes and formulas is defined inductively on the structure of formulas.

$E \models \mathtt{tt}$		
$E \not\models \mathtt{ff}$		
$E \models \Phi \wedge \Psi$	iff	$E \models \Phi$ and $E \models \Psi$
$E \models \Phi \vee \Psi$	iff	$E \models \Phi$ or $E \models \Psi$
$E \models [K]\Phi$	iff	$\forall F \in \{E' : E \xrightarrow{a} E'$ and $a \in K\}.\ F \models \Phi$
$E \models \langle K \rangle \Phi$	iff	$\exists F \in \{E' : E \xrightarrow{a} E'$ and $a \in K\}.\ F \models \Phi$

Every process has the property tt, whereas no process has the property ff. A process has the property $\Phi \wedge \Psi$ when it has the property Φ and the property Ψ, and it satisfies $\Phi \vee \Psi$ if it satisfies one of the disjuncts. The definition of satisfaction between processes and formulas prefaced by a modal operator appeals to behaviour of processes as given by the rules for transitions. A process E has the property $[K]\Phi$ if every process which E evolves to after carrying out any action in K has the property Φ. And E satisfies $\langle K \rangle \Phi$ if E can become a process that satisfies Φ by carrying out an action in K. To reduce the number of brackets in modalities, we write $[a_1, \ldots, a_n]$ and $\langle a_1, \ldots, a_n \rangle$ instead of $[\{a_1, \ldots, a_n\}]$ and $\langle \{a_1, \ldots, a_n\} \rangle$. The modal logic M slightly generalises Hennessy-Milner logic, due to Hennessy and Milner [29], because sets of actions, instead of single actions, occur in the modalities.

The simple formula $\langle \mathtt{tick} \rangle \mathtt{tt}$ expresses an ability to carry out the action tick. Process E has this property provided that there is a transition $E \xrightarrow{\mathtt{tick}} F$. The clock Cl from Section 1.1 has this property, whereas the vending machine Ven of Figure 1.2 does not. In contrast, the formula $[\mathtt{tick}]\mathtt{ff}$ expresses an inability to carry out the action tick, because process E satisfies $[\mathtt{tick}]\mathtt{ff}$ if E does not have a transition $E \xrightarrow{\mathtt{tick}} F$. These basic properties can be embedded within modal operators and between boolean connectives. An example is the formula $[\mathtt{tick}](\langle\mathtt{tick}\rangle\mathtt{tt} \wedge [\mathtt{tock}]\mathtt{ff})$[1], which expresses that, after any tick action, it is possible to perform tick again, but not possible to perform tock.

Example 1 Cl has the property $[\mathtt{tick}](\langle\mathtt{tick}\rangle\mathtt{tt} \wedge [\mathtt{tock}]\mathtt{ff})$. Applying the definition of satisfaction Cl has this property:

$$\begin{array}{ll}
\text{iff} & \forall F \in \{E : \mathtt{Cl} \stackrel{\mathtt{tick}}{\longrightarrow} E\}.\ F \models \langle\mathtt{tick}\rangle\mathtt{tt} \wedge [\mathtt{tock}]\mathtt{ff} \\
\text{iff} & \mathtt{Cl} \models \langle\mathtt{tick}\rangle\mathtt{tt} \wedge [\mathtt{tock}]\mathtt{ff} \\
\text{iff} & \mathtt{Cl} \models \langle\mathtt{tick}\rangle\mathtt{tt} \text{ and } \mathtt{Cl} \models [\mathtt{tock}]\mathtt{ff} \\
\text{iff} & \exists F \in \{E : \mathtt{Cl} \stackrel{\mathtt{tick}}{\longrightarrow} E\} \text{ and } \mathtt{Cl} \models [\mathtt{tock}]\mathtt{ff} \\
\text{iff} & \exists F \in \{\mathtt{Cl}\} \text{ and } \mathtt{Cl} \models [\mathtt{tock}]\mathtt{ff} \\
\text{iff} & \mathtt{Cl} \models [\mathtt{tock}]\mathtt{ff} \\
\text{iff} & \{E : \mathtt{Cl} \stackrel{\mathtt{tock}}{\longrightarrow} E\} = \emptyset \\
\text{iff} & \emptyset = \emptyset
\end{array}$$

On the other hand, $\mathtt{Cl} \not\models [\mathtt{tick}](\langle\mathtt{tock}\rangle\mathtt{tt} \vee [\mathtt{tick}]\mathtt{ff})$.

The formula $\langle K\rangle\mathtt{tt}$ expresses a capability for carrying out some action in K, whereas $[K]\mathtt{ff}$ expresses an inability to initially perform any action in K. In the case of the vending machine Ven, a button cannot be depressed before money is deposited, so $\mathtt{Ven} \models [\mathtt{big}, \mathtt{little}]\mathtt{ff}$. Other interesting properties of Ven are as follows.

- $\mathtt{Ven} \models [\mathtt{2p}]([\mathtt{little}]\mathtt{ff} \wedge \langle\mathtt{big}\rangle\mathtt{tt})$: after 2p is deposited the little button cannot be depressed whereas the big one can
- $\mathtt{Ven} \models [\mathtt{1p}, \mathtt{2p}][\mathtt{1p}, \mathtt{2p}]\mathtt{ff}$: after a coin is entrusted no other coin (2p or 1p) may be deposited
- $\mathtt{Ven} \models [\mathtt{1p}, \mathtt{2p}][\mathtt{big}, \mathtt{little}]\langle\mathtt{collect_b}, \mathtt{collect_l}\rangle\mathtt{tt}$: after a coin is deposited and a button is depressed, an item can be collected

Verifying that Ven has these properties is undemanding. A proof merely appeals to the inductive definition of the satisfaction relation between a process and a formula, which may rely on the rules for transitions. Similarly, establishing that a process lacks a property is equally routine.

[1] We assume that $\wedge$ and $\vee$ have wider scope than the modalities $[K]$, $\langle K\rangle$, and that brackets are introduced to resolve any further ambiguities as to the structure of a formula. Therefore, $\wedge$ is the main connective of the subformula $\langle\mathtt{tick}\rangle\mathtt{tt} \wedge [\mathtt{tock}]\mathtt{ff}$.

Example 2 $\texttt{Ven} \not\models \langle \texttt{1p} \rangle \langle \texttt{1p}, \texttt{big} \rangle \texttt{tt}$

$$
\begin{aligned}
&\text{iff} \quad \text{not}\,(\exists F \in \{E : \texttt{Ven} \xrightarrow{\texttt{1p}} E\}.\ F \models \langle \texttt{1p}, \texttt{big} \rangle \texttt{tt}) \\
&\text{iff} \quad \texttt{Ven}_1 \not\models \langle \texttt{1p}, \texttt{big} \rangle \texttt{tt} \\
&\text{iff} \quad \text{not}\,(\exists F \in \{E : \texttt{Ven}_1 \xrightarrow{a} E \text{ and } a \in \{\texttt{1p}, \texttt{big}\}\}) \\
&\text{iff} \quad \{E : \texttt{Ven}_1 \xrightarrow{a} E \text{ and } a \in \{\texttt{1p}, \texttt{big}\}\} = \emptyset \\
&\text{iff} \quad \emptyset = \emptyset
\end{aligned}
$$

Again, this demonstration appeals to the inductive definition of satisfaction between a process and a formula.

When showing that a process has, or fails to have, a property we do not need to build its transition graph, as the following example illustrates.

Example 3 Consider the following features of the crossing of Section 1.2.

$$
\begin{aligned}
&\texttt{Crossing} \models [\texttt{train}](\langle \tau \rangle \texttt{tt} \wedge \langle \texttt{car} \rangle [\tau][\tau]\texttt{ff}) \\
&\texttt{Crossing} \models [\texttt{car}][\texttt{train}][\tau](\langle \overline{\texttt{tcross}} \rangle \texttt{tt} \vee \langle \overline{\texttt{ccross}} \rangle \texttt{tt}) \\
&\texttt{Crossing} \not\models [\texttt{car}][\texttt{train}][\tau](\langle \overline{\texttt{tcross}} \rangle \texttt{tt} \wedge \langle \overline{\texttt{ccross}} \rangle \texttt{tt})
\end{aligned}
$$

Proofs of these depend only on part of the behaviour of the crossing. For instance, the first relies only on the processes E_2, E_3, E_5, E_6 and E_7 of Figure 1.12.

Actions in the modalities may contain values. For instance, the register $\texttt{Reg}_5$ from Section 1.1 can only transmit the value 5, whereas it can be overwritten by any value $k \geq 0$.

$$
\begin{aligned}
&\texttt{Reg}_5 \models \langle \overline{\texttt{read}}(5) \rangle \texttt{tt} \wedge [\{\overline{\texttt{read}}(n) : n \neq 5\}]\texttt{ff} \\
&\texttt{Reg}_5 \models \langle \texttt{write}(k) \rangle \texttt{tt}
\end{aligned}
$$

Assume that A is a universal set of actions including τ. Hence, A is the set $\mathsf{O} \cup \{\tau\}$, where O is the general set of observable actions described in Section 1.3. A little notation is now introduced for sets of actions within modalities. The set $-K$ abbreviates $\mathsf{A} - K$, and $-a_1, \ldots, a_n$ abbreviates $-\{a_1, \ldots, a_n\}$. Also, assume that $-$ abbreviates the set $-\emptyset$, which is therefore the set A. A process E has the property $[-]\Phi$ when each member of the set $\{E' : E \xrightarrow{a} E' \text{ and } a \in \mathsf{A}\}$ satisfies Φ. The modal formula $[-]\texttt{ff}$ therefore expresses deadlock or termination, an inability to carry out any action whatever.

Within M, one can express immediate "necessity" or "inevitability." The property "a must happen next" is given by the formula $\langle - \rangle \texttt{tt} \wedge [-a]\texttt{ff}$. The conjunct $\langle - \rangle \texttt{tt}$ affirms that an action is possible, whereas $[-a]\texttt{ff}$ states that every action except a is impossible. After 2p is deposited, Ven must perform big, and so $\texttt{Ven} \models [\texttt{2p}](\langle - \rangle \texttt{tt} \wedge [-\texttt{big}]\texttt{ff})$. Ven also has the following property, that the third action it performs must be a collect.

$$
\langle - \rangle \texttt{tt} \wedge [-](\langle - \rangle \texttt{tt} \wedge [-](\langle - \rangle \texttt{tt} \wedge [-\texttt{collect}_1, \texttt{collect}_b]\texttt{ff}))
$$

Exercises

1. Show the following

   a. $\texttt{Ven} \models [\texttt{2p}, \texttt{1p}]\langle\texttt{big}, \texttt{little}\rangle\texttt{tt}$
   b. $\texttt{Ven} \not\models [\texttt{2p}, \texttt{1p}](\langle\texttt{big}\rangle\texttt{tt} \wedge \langle\texttt{little}\rangle\texttt{tt})$
   c. $\texttt{Ven} \models [\texttt{2p}]([-\texttt{big}]\texttt{ff} \wedge \langle-\texttt{little}, \texttt{2p}\rangle\texttt{tt})$.
   d. $\texttt{Cnt} \models [\texttt{up}]\langle\texttt{down}\rangle[\texttt{down}]\texttt{ff}$
   e. $\texttt{Crossing} \models [\texttt{train}](\langle\tau\rangle\texttt{tt} \wedge \langle\texttt{car}\rangle[\tau][\tau]\texttt{ff})$
   f. $\texttt{Crossing} \models [\texttt{car}][\texttt{train}][\tau](\langle\overline{\texttt{tcross}}\rangle\texttt{tt} \vee \langle\overline{\texttt{ccross}}\rangle\texttt{tt})$
   g. $\texttt{Crossing} \not\models [\texttt{car}][\texttt{train}][\tau](\langle\overline{\texttt{tcross}}\rangle\texttt{tt} \wedge \langle\overline{\texttt{ccross}}\rangle\texttt{tt})$

   where Cnt and Crossing are defined in Section 1.2.

2. Show $\texttt{Ct}_3 \models \langle\texttt{down}\rangle\langle\texttt{up}\rangle\langle\texttt{down}\rangle\langle\texttt{down}\rangle\texttt{tt}$, but that $\texttt{Ct}_1$ fails to have this property when $\texttt{Ct}_i$ is from Figure 1.4. Using induction, show that, for any i and j, $\texttt{Ct}_i \models [\texttt{up}]^j\langle\texttt{down}\rangle^j\texttt{tt}$, that whatever goes up may come down in equal proportions, where $[a]^0\Phi = \Phi$ and $[a]^{n+1}\Phi = [a][a]^n\Phi$ (and similarly for $\langle a\rangle^n\Phi$).

3. Using induction on i show

   $$\texttt{Cop}' \models [\texttt{no}(i)][\texttt{in}(v)]\langle\overline{\texttt{out}}(v)\rangle^{i+1}\texttt{tt}$$

   where $\texttt{Cop}'$ is defined in Section 1.1.

4. Consider the following three vending machines.

   $$\begin{aligned} \texttt{Ven}_1 &\stackrel{\text{def}}{=} \texttt{1p.1p.(tea.Ven}_1 + \texttt{coffee.Ven}_1) \\ \texttt{Ven}_2 &\stackrel{\text{def}}{=} \texttt{1p.(1p.tea.Ven}_2 + \texttt{1p.coffee.Ven}_2) \\ \texttt{Ven}_3 &\stackrel{\text{def}}{=} \texttt{1p.1p.tea.Ven}_3 + \texttt{1p.1p.coffee.Ven}_3 \end{aligned}$$

   Give modal formulas that distinguish between them: that is, find formulas Φ_j, $1 \leq j \leq 3$, such that $\texttt{Ven}_j \models \Phi_j$ but $\texttt{Ven}_i \not\models \Phi_j$ when $i \neq j$.

5. Let $\texttt{Cl} \stackrel{\text{def}}{=} \texttt{tick.Cl}$ and $\texttt{Cl}_2 \stackrel{\text{def}}{=} \texttt{tick.tick.Cl}_2$. Show that no modal formula distinguishes between these clocks. That is, prove that $\texttt{Cl} \models \Phi$ iff $\texttt{Cl}_2 \models \Phi$ for all modal formulas Φ.

6. A modal formula Φ distinguishes between two processes E and F if either $E \models \Phi$ and $F \not\models \Phi$, or $E \not\models \Phi$ and $F \models \Phi$. Provide a modal formula that distinguishes between $\texttt{Sched}_4$ and $\texttt{Sched}_4'$ of Section 1.4.

7. Express as a modal formula the property "the second action must be a (parameterised) $\overline{\texttt{out}}$ action," and show that Cop of Section 1.1 has this property.

8. Express as a modal formula the property "the fourth action must be τ," and show that the slot machine $\texttt{SM}_n$ of Figure 1.15 has this property.

2.2 Hennessy-Milner logic II

The modal logic M, as presented in the previous section, does not contain a negation operator $\neg$. The semantic clause for negation is as follows.

$$E \models \neg\Phi \text{ iff } E \not\models \Phi$$

However, for any formula Φ of M, there is the formula Φ^c that expresses the negation of Φ. The complementation operator ^{c} is defined inductively as follows.

$$\begin{array}{llllll} \mathtt{tt}^c & = & \mathtt{ff} & \mathtt{ff}^c & = & \mathtt{tt} \\ (\Phi_1 \wedge \Phi_2)^c & = & \Phi_1^c \vee \Phi_2^c & (\Phi_1 \vee \Phi_2)^c & = & \Phi_1^c \wedge \Phi_2^c \\ ([K]\Phi)^c & = & \langle K\rangle\Phi^c & (\langle K\rangle\Phi)^c & = & [K]\Phi^c \end{array}$$

Φ^c is the result of replacing each operator in Φ with its "dual," where $\mathtt{tt}$ and $\mathtt{ff}$, $\wedge$ and $\vee$, and $[K]$ and $\langle K\rangle$ are duals. For instance, the complement of $[\mathtt{tick}](\langle\mathtt{tick}\rangle\mathtt{tt} \wedge [\mathtt{tock}]\mathtt{ff})$ is the formula $\langle\mathtt{tick}\rangle([\mathtt{tick}]\mathtt{ff} \vee \langle\mathtt{tock}\rangle\mathtt{tt})$. The following result shows that Φ^c expresses $\neg\Phi$.

Proposition 1 $E \models \Phi^c$ *iff* $E \not\models \Phi$.

Proof. By induction on the structure of Φ, we show that, for any process F, $F \models \Phi^c$ iff $F \not\models \Phi$. The base cases are when $\Phi = \mathtt{tt}$ and $\Phi = \mathtt{ff}$. Clearly, $F \models \mathtt{ff}$ iff $F \not\models \mathtt{tt}$ and $F \models \mathtt{tt}$ iff $F \not\models \mathtt{ff}$. For the induction step, assume the result for formulas Φ_1 and Φ_2. If we can show that it also holds for $\Phi_1 \wedge \Phi_2$, $\Phi_1 \vee \Phi_2$, $[K]\Phi_1$ and $\langle K\rangle\Phi_1$, then the result is proved. Let $\Phi = \Phi_1 \wedge \Phi_2$. $F \models (\Phi_1 \wedge \Phi_2)^c$

$$\begin{array}{ll} \text{iff} & F \models \Phi_1^c \vee \Phi_2^c \ \text{(by definition of } ^c) \\ \text{iff} & F \models \Phi_1^c \text{ or } F \models \Phi_2^c \ \text{(by clause for } \vee) \\ \text{iff} & F \not\models \Phi_1 \text{ or } F \not\models \Phi_2 \ \text{(by induction hypothesis)} \\ \text{iff} & F \not\models \Phi_1 \wedge \Phi_2 \ \text{(by clause for } \wedge). \end{array}$$

The case $\Phi = \Phi_1 \vee \Phi_2$ is very similar. Let $\Phi = [K]\Phi_1$. $F \models ([K]\Phi_1)^c$

$$\begin{array}{ll} \text{iff} & F \models \langle K\rangle\Phi_1^c \ \text{(by definition of } ^c) \\ \text{iff} & \exists G. \exists a \in K. F \stackrel{a}{\longrightarrow} G \text{ and } G \models \Phi_1^c \ \text{(by clause for } \langle K\rangle) \\ \text{iff} & \exists G. \exists a \in K. F \stackrel{a}{\longrightarrow} G \text{ and } G \not\models \Phi_1 \ \text{(by induction hypothesis)} \\ \text{iff} & F \not\models [K]\Phi_1 \ \text{(by clause for } [K]). \end{array}$$

The final case $\Phi = \langle K\rangle\Phi_1$ is similar. □

To show that a process fails to have a property is therefore equivalent to showing that it has the complement property. Notice that the complement of a complement of a formula is the formula itself, $(\Phi^c)^c = \Phi$.

In Section 1.6 we defined the set of subprocesses Sub(E) of a process E. This set may have infinite size. We now inductively define the set of subformulas,

$\mathrm{Sub}(\Phi)$, of a formula Φ.

$$
\begin{array}{lcl}
\mathrm{Sub}(\mathtt{tt}) & = & \{\mathtt{tt}\} \\
\mathrm{Sub}(\mathtt{ff}) & = & \{\mathtt{ff}\} \\
\mathrm{Sub}(\Phi_1 \wedge \Phi_2) & = & \{\Phi_1 \wedge \Phi_2\} \cup \mathrm{Sub}(\Phi_1) \cup \mathrm{Sub}(\Phi_2) \\
\mathrm{Sub}(\Phi_1 \vee \Phi_2) & = & \{\Phi_1 \vee \Phi_2\} \cup \mathrm{Sub}(\Phi_1) \cup \mathrm{Sub}(\Phi_2) \\
\mathrm{Sub}([K]\Phi) & = & \{[K]\Phi\} \cup \mathrm{Sub}(\Phi) \\
\mathrm{Sub}(\langle K\rangle\Phi) & = & \{\langle K\rangle\Phi\} \cup \mathrm{Sub}(\Phi)
\end{array}
$$

For any formula Φ, $\mathrm{Sub}(\Phi)$ is a finite set of formulas. For instance, if Φ is the formula $([\mathtt{tick}](\langle\mathtt{tick}\rangle\mathtt{tt} \wedge [\mathtt{tock}]\mathtt{ff}))$, then $\mathrm{Sub}(\Phi)$ is the following set.

$$\{\Phi, \langle\mathtt{tick}\rangle\mathtt{tt} \wedge [\mathtt{tock}]\mathtt{ff}, \langle\mathtt{tick}\rangle\mathtt{tt}, [\mathtt{tock}]\mathtt{ff}, \mathtt{tt}, \mathtt{ff}\}$$

The size of a modal formula Ψ, denoted by $|\Psi|$, is the number of occurrences of $\mathtt{tt}$, $\mathtt{ff}$, $\wedge$, $\vee$, $[K]$ and $\langle K\rangle$ within it. Clearly, the number of formulas in the set $\mathrm{Sub}(\Phi)$ is no more than $|\Phi|$.

A modal formula is "realizable" (or "satisfiable") if there is a process that satisfies it. $[\mathtt{tick}](\langle\mathtt{tick}\rangle\mathtt{tt} \wedge [\mathtt{tock}]\mathtt{ff})$ is realizable because $\mathtt{Cl}$ satisfies it. On the other hand, $\langle\mathtt{tick}\rangle(\langle\mathtt{tick}\rangle\mathtt{tt} \wedge [\mathtt{tick}]\mathtt{ff})$ is not realizable because a process cannot tick, and then be able to both tick again and fail to tick. There is a simple technique for determining whether a formula is realizable, provided it does not contain modalities with values. First, realizability is extended to finite sets of formulas: the finite set Γ is realizable if there is a process satisfying every Φ in Γ. The method for deciding realizability of a set of formulas consists of reducing it to realizability of smaller sized sets, by stripping away connectives. The size of a set is the sum of the sizes of its formulas. An example reduction is that $\Gamma \cup \{\Phi \wedge \Psi\}$ becomes the smaller set $\Gamma \cup \{\Phi, \Psi\}$. The details are left as an exercise for the reader.

The presence of values in modal operators suggests that a more extensive modal logic is appropriate to property expression that permits generality of value. One extension is to include first-order quantification over values. For instance $\mathtt{Reg}_k$ has the property $\forall n.\,\langle\mathtt{write}(n)\rangle\mathtt{tt}$ where n in the modal operator is bound by the universal quantifier: here it is implicit that n ranges over $\mathbb{N}$. Quantifiers allow value dependence to be directly expressible. Cop has the property $\forall d.\,[\mathtt{in}(d)](\langle\overline{\mathtt{out}}(d)\rangle\mathtt{tt}\wedge[-\overline{\mathtt{out}}(d)]\mathtt{ff})$, where d ranges over the appropriate value space D. Semantic clauses for the quantifiers are as follows:

$$
\begin{array}{lll}
E \models \forall x.\,\Phi & \text{iff} & \forall d \in D.\ E \models \Phi\{d/x\} \\
E \models \exists x.\,\Phi & \text{iff} & \exists d \in D.\ E \models \Phi\{d/x\},
\end{array}
$$

where $\{d/x\}$ is substitution of d for all free occurrences of x. Predicates over values can also be included (for example, expressing evenness of an integer). This leads to very rich first-order modal logics. We leave the reader to spell out some of the possibilities here.

An alternative to using quantifiers over values consists of using infinite conjunction and disjunction. Infinitary modal logic M_∞, where I ranges over arbitrary finite and infinite indexing families, is defined as follows.

$$\Phi ::= \bigwedge\{\Phi_i \,:\, i \in I\} \mid \bigvee\{\Phi_i \,:\, i \in I\} \mid [K]\Phi \mid \langle K\rangle\Phi$$

The satisfaction relation between processes and $\bigwedge$ and $\bigvee$ formulas is defined below.

$$E \models \bigwedge\{\Phi_i \,:\, i \in I\} \quad \text{iff} \quad E \models \Phi_j \text{ for every } j \in I$$
$$E \models \bigvee\{\Phi_i \,:\, i \in I\} \quad \text{iff} \quad E \models \Phi_j \text{ for some } j \in I$$

$\mathtt{tt}$ abbreviates $\bigwedge\{\Phi_i \,:\, i \in \emptyset\}$ and $\mathtt{ff}$ abbreviates $\bigvee\{\Phi_i \,:\, i \in \emptyset\}$. Quantified formulas can be interpreted in M_∞. For instance,

$$\forall d.\, [\mathtt{in}(d)](\langle\overline{\mathtt{out}}(d)\rangle\mathtt{tt} \wedge [-\overline{\mathtt{out}}(d)]\mathtt{ff})$$

can be expressed as

$$\bigwedge\{[\mathtt{in}(d)](\langle\overline{\mathtt{out}}(d)\rangle\mathtt{tt} \wedge [-\overline{\mathtt{out}}(d)]\mathtt{ff}) \,:\, d \in D\}.$$

Exercises

1. For each of the following formulas, determine its complement.
   - **a.** $\langle a_1\rangle\langle a_2\rangle\langle a_3\rangle\mathtt{tt}$
   - **b.** $\langle a_1\rangle\langle a_2\rangle\langle a_3\rangle[-]\mathtt{ff}$
   - **c.** $[\mathtt{train}](\langle\tau\rangle\mathtt{tt} \wedge \langle\mathtt{car}\rangle[\tau][\tau]\mathtt{ff})$
   - **d.** $\langle\overline{\mathtt{read}}(5)\rangle\mathtt{tt} \wedge [\{\overline{\mathtt{read}}(n) \,:\, n \neq 55\}]\mathtt{ff}$
   - **e.** $\langle -\rangle\mathtt{tt} \wedge [-](\langle -\rangle\mathtt{tt} \wedge [-](\langle -\rangle\mathtt{tt} \wedge [-\mathtt{collect}_1, \mathtt{collect}_b]\mathtt{ff}))$
2. Prove that $(\Phi^c)^c = \Phi$.
3. For each of the following, determine whether it is realizable, and when it is, exhibit a realizer.
   - **a.** $\langle\mathtt{tick}\rangle[\mathtt{tock}](\langle\mathtt{tick}\rangle\mathtt{tt} \wedge [\mathtt{tick}]\mathtt{ff})$
   - **b.** $\langle\mathtt{tick}\rangle[\mathtt{tock}](\langle\mathtt{tick}\rangle\mathtt{tt} \wedge [\mathtt{tick}]\mathtt{ff}) \wedge [-]\langle\mathtt{tock}\rangle\mathtt{tt}$
   - **c.** $\langle\mathtt{tick}\rangle[\mathtt{tock}](\langle\mathtt{tick}\rangle\mathtt{tt} \wedge [\mathtt{tick}]\mathtt{ff}) \wedge [-]\langle -\rangle\mathtt{tt}$
   - **d.** $[-]([-](\langle -\rangle\mathtt{tt} \wedge [-\mathtt{collect}_1, \mathtt{collect}_b]\mathtt{ff}) \wedge \langle -\rangle\mathtt{tt}) \wedge \langle -\rangle\mathtt{tt}$
   - **e.** $[\mathtt{in}(5)](\langle\overline{\mathtt{out}}(5)\rangle\mathtt{tt} \wedge \langle\overline{\mathtt{out}}(7)\rangle\mathtt{tt})$
4. Design an algorithm that decides whether a modal formula of M is realisable.
5. A modal formula is *valid* if every process satisfies it. Show that Φ is valid iff Φ^c is not realizable. Let $\rightarrow$ be the implies connective whose definition is

   $$\Phi \rightarrow \Psi \stackrel{\text{def}}{=} \Phi^c \vee \Psi.$$

   Which of the following are valid when Φ and Ψ are arbitrary modal formulas?
   - **a.** $\langle\mathtt{tick}\rangle(\Phi \vee \Psi) \rightarrow (\langle\mathtt{tick}\rangle\Phi \vee \langle\mathtt{tick}\rangle\Psi)$

b. $(\langle \texttt{tick} \rangle \Phi \wedge \langle \texttt{tick} \rangle \Psi) \rightarrow \langle \texttt{tick} \rangle (\Phi \wedge \Psi)$

c. $[\texttt{tick}](\Phi \rightarrow \Psi) \rightarrow ([\texttt{tick}]\Phi \rightarrow [\texttt{tick}]\Psi)$

d. $([\texttt{tick}]\Phi \rightarrow [\texttt{tick}]\Psi) \rightarrow [\texttt{tick}](\Phi \rightarrow \Psi)$

6. Two modal formulas Φ and Ψ are "equivalent" if, for all processes E, $E \models \Phi$ iff $E \models \Psi$. Which of the following pairs are equivalent when Φ_1 and Φ_2 are arbitrary modal formulas?

a. $\langle \texttt{tick} \rangle (\Phi_1 \wedge \Phi_2), \quad \langle \texttt{tick} \rangle \Phi_1 \wedge \langle \texttt{tick} \rangle \Phi_2$

b. $\langle \texttt{tick} \rangle (\Phi_1 \vee \Phi_2), \quad \langle \texttt{tick} \rangle \Phi_1 \vee \langle \texttt{tick} \rangle \Phi_2$

c. $[\texttt{tick}](\Phi_1 \wedge \Phi_2), \quad [\texttt{tick}]\Phi_1 \wedge [\texttt{tick}]\Phi_2$

d. $[\texttt{tick}](\Phi_1 \vee \Phi_2), \quad [\texttt{tick}]\Phi_1 \vee [\texttt{tick}]\Phi_2$

e. $[\texttt{tick}]\Phi_1, \quad \langle \texttt{tick} \rangle \Phi_1$

7. Define first-order modal logic for value-passing processes, where the quantifiers range over values.

2.3 Algebraic structure and modal properties

Process behaviour is chronicled through transitions. Processes also have structure, defined as they are from combinators. An interesting issue is the extent to which properties of processes are definable from this structure, without appealing to transitional behaviour. The ascription of boolean combinations of properties to processes does not immediately depend on their behaviour. For instance, E satisfies $\Phi \vee \Psi$ if and only if E satisfies one of the disjuncts. Therefore, it is the modal operators that we need to concern ourselves with, and how algebraic structure relates to them. A variety of cases is covered in the following proposition.

Proposition 1

1. *If* $a \notin K$ *then* $a.E \models [K]\Phi$ *and* $a.E \not\models \langle K \rangle \Phi$
2. *If* $a \in K$ *then* $a.E \models [K]\Phi$ *iff* $E \models \Phi$
3. *If* $a \in K$ *then* $a.E \models \langle K \rangle \Phi$ *iff* $E \models \Phi$
4. $\sum\{E_i : i \in I\} \models [K]\Phi$ *iff for all* $j \in I.\ E_j \models [K]\Phi$
5. $\sum\{E_i : i \in I\} \models \langle K \rangle \Phi$ *iff for some* $j \in I.\ E_j \models \langle K \rangle \Phi$
6. *If* $P \stackrel{\text{def}}{=} E$ *and* $E \models \Phi$ *then* $P \models \Phi$

Proof. For 1, if $a \notin K$, then the set $\{F : a.E \stackrel{b}{\longrightarrow} F \text{ and } b \in K\} = \emptyset$, and so $a.E \models [K]\Phi$, but $a.E \not\models \langle K \rangle \Phi$. Cases 2 and 3 follow from the observation if $a \in K$, then the set $\{F : a.E \stackrel{b}{\longrightarrow} F \text{ and } b \in K\} = \{E\}$. 4 and 5 depend on the following equality, $\{F : \sum\{E_i : i \in I\} \stackrel{a}{\longrightarrow} F \text{ and } a \in K\}$ is the same set as

$\{F : E_j \xrightarrow{a} F$ and $a \in K$ and $j \in I\}$. For 6, observe that the behaviour of P when $P \stackrel{\text{def}}{=} E$ is that of E. $\square$

Example 1 Using Proposition 1 and the semantic clauses for boolean combinations of properties, we can now show that the vending machine $\mathtt{Ven}$ of Figure 1.2 has the property $[2p](\langle -\rangle \mathtt{tt} \wedge [-\mathtt{big}]\mathtt{ff})$ without appealing to transitional behaviour. Using Proposition 1.6, this is established if

$$\mathtt{2p.Ven_b} + \mathtt{1p.Ven_l} \models [2p](\langle -\rangle \mathtt{tt} \wedge [-\mathtt{big}]\mathtt{ff}),$$

which reduces by 1.4 to demonstrating

$$\mathtt{2p.Ven_b} \models [2p](\langle -\rangle \mathtt{tt} \wedge [-\mathtt{big}]\mathtt{ff}) \text{ and}$$
$$\mathtt{1p.Ven_l} \models [2p](\langle -\rangle \mathtt{tt} \wedge [-\mathtt{big}]\mathtt{ff}).$$

The second follows from Proposition 1.1 because $\mathtt{1p} \notin \{2p\}$. Using Proposition 1.2, the first reduces to showing

$$\mathtt{Ven_b} \models \langle -\rangle \mathtt{tt} \wedge [-\mathtt{big}]\mathtt{ff}.$$

By Proposition 1.6, this is established if

$$\mathtt{big.collect_b.Ven} \models \langle -\rangle \mathtt{tt} \wedge [-\mathtt{big}]\mathtt{ff}.$$

That is, if the following pair hold

$$\mathtt{big.collect_b.Ven} \models \langle -\rangle \mathtt{tt} \text{ and}$$
$$\mathtt{big.collect_b.Ven} \models [-\mathtt{big}]\mathtt{ff}.$$

The second of these is true by Proposition 1.1, and the first is established using Proposition 1.2 because $\mathtt{collect_b.Ven} \models \mathtt{tt}$.

The effect of removing the restriction $\backslash J$ from a process can be captured by inductively defining an operator on modal formulas, $\Phi\backslash J$. The intention is that $E\backslash J \models \Phi$ iff $E \models \Phi\backslash J$. In the following, let J^+ be the set $J \cup \overline{J}$.

$$\begin{array}{llllll} \mathtt{tt}\backslash J & = & \mathtt{tt} & \mathtt{ff}\backslash J & = & \mathtt{ff} \\ (\Phi \wedge \Psi)\backslash J & = & \Phi\backslash J \wedge \Psi\backslash J & (\Phi \vee \Psi)\backslash J & = & \Phi\backslash J \vee \Psi\backslash J \\ ([K]\Phi)\backslash J & = & [K - J^+](\Phi\backslash J) & (\langle K\rangle\Phi)\backslash J & = & \langle K - J^+\rangle(\Phi\backslash J) \end{array}$$

The operator $\backslash J$ removes actions from modalities, as the following example illustrates.

$$[\mathtt{tick}, \mathtt{tock}](\langle -\rangle \mathtt{tt} \wedge [\mathtt{tock}]\mathtt{ff})\backslash\{\mathtt{tick}\} = [\mathtt{tock}](\langle -\mathtt{tick}\rangle \mathtt{tt} \wedge [\mathtt{tock}]\mathtt{ff})$$

The operator $\backslash J$ on formulas is an inverse of its application to processes, as the next result shows.

Proposition 2 *$E\backslash J \models \Phi$ iff $E \models \Phi\backslash J$.*

Proof. By induction on Φ. If Φ is $\mathtt{tt}$ or $\mathtt{ff}$, then the result is clear. Suppose Φ is $\Phi_1 \wedge \Phi_2$. So, $E\backslash J \models \Phi$ iff $E\backslash J \models \Phi_1$ and $E\backslash J \models \Phi_2$ iff by the induction hypothesis $E \models \Phi_1\backslash J$ and $E \models \Phi_2\backslash J$, and now by the inductive definition above iff $E \models \Phi\backslash J$. The case when Φ is $\Phi_1 \vee \Phi_2$ is similar. Assume Φ is $[K]\Psi$ and $E\backslash J \models [K]\Psi$, but $E \not\models [K - J^+](\Psi\backslash J)$. Therefore, $E \xrightarrow{a} F$ with $a \in K - J^+$ and $F \not\models \Psi\backslash J$. By the induction hypothesis it follows that $F\backslash J \not\models \Psi$, and because $a \in K - J^+$ we also know that $E\backslash J \xrightarrow{a} F\backslash J$. But then we have a contradiction because this shows that $E\backslash J \not\models [K]\Psi$. For the other direction, suppose that $E \models [K - J^+](\Psi\backslash J)$, but $E\backslash J \not\models [K]\Psi$. Therefore, $E\backslash J \xrightarrow{a} F\backslash J$ for some $a \in K$ and $F\backslash J \not\models \Psi$. But this a must belong to $K - J^+$ by the transition rule for $\backslash J$, and by the induction hypothesis $F \not\models \Psi\backslash J$. Therefore, $E \not\models ([K]\Psi)\backslash J$. The final case when Φ is $\langle K\rangle\Psi$ is similar. □

There are similar inverse operations on formulas for renaming $[f]$ of Section 1.4 and for hiding $\backslash\backslash J$ of Section 1.5. We leave their exact definition as an exercise.

Much more troublesome is coping with parallel composition. One idea, which is not entirely satisfactory, is to define an "inverse" of parallel on formulas. For each process F, and for each formula Φ, one defines the new formula Φ/F with the intention that for any E, $E \mid F \models \Phi$ iff $E \models \Phi/F$. Instead of presenting an inductive definition of this "slicing operator," Φ/F, we illustrate a particular use of it.

Example 2 Let Ven be the vending machine and consider the following.

$$\begin{array}{lcl} \mathtt{Use}_1 & \stackrel{\text{def}}{=} & \overline{\mathtt{1p}}.\overline{\mathtt{little}}.\mathtt{Use}_1 \\ K & = & \{\mathtt{1p}, \mathtt{2p}, \mathtt{little}, \mathtt{big}\} \\ K^+ & = & K \cup \overline{K} \end{array}$$

We show that $(\mathtt{Ven} \mid \mathtt{Use}_1)\backslash K \models [-][-]\langle\mathtt{collect}_1\rangle\mathtt{tt}$. Proposition 2 is applied first.

$$\begin{array}{rcll} \mathtt{Ven} \mid \mathtt{Use}_1 & \models & ([-][-]\langle\mathtt{collect}_1\rangle\mathtt{tt})\backslash K & \text{iff} \\ \mathtt{Ven} \mid \mathtt{Use}_1 & \models & [-K^+][-K^+]\langle\mathtt{collect}_1\rangle\mathtt{tt} & \text{iff} \\ \mathtt{Ven} & \models & ([-K^+][-K^+]\langle\mathtt{collect}_1\rangle\mathtt{tt})/\mathtt{Use}_1 & \end{array}$$

We need to understand the formula $([-K^+][-K^+]\langle\mathtt{collect}_1\rangle\mathtt{tt})/\mathtt{Use}_1$. It is the same as $([-K^+][-K^+]\langle\mathtt{collect}_1\rangle\mathtt{tt})/\overline{\mathtt{1p}}.\overline{\mathtt{little}}.\mathtt{Use}_1$. We want to distribute the process through the formula. The action $\overline{\mathtt{1p}}$ can not directly contribute to the modality $[-K^+]$ because $\overline{\mathtt{1p}} \in K^+$. However, it can contribute as part of a communication if Ven has 1p transitions. Therefore, either Ven contributes a transition, or there is a communication between Ven and the user. Therefore, we need to show the following pair.

$$\mathtt{Ven} \models [-K^+]([-K^+]\langle\mathtt{collect}_1\rangle\mathtt{tt}/\overline{\mathtt{1p}}.\overline{\mathtt{little}}.\mathtt{Use}_1)$$
$$\mathtt{Ven} \models [\mathtt{1p}]([-K^+]\langle\mathtt{collect}_1\rangle\mathtt{tt}/\overline{\mathtt{little}}.\mathtt{Use}_1)$$

The first of these is derivable using Proposition 1. By the same Proposition, the second is quivalent to the following.

$$\mathtt{Ven_1} \models ([-K^+]\langle\mathtt{collect_1}\rangle\mathtt{tt})/\overline{\mathtt{little}}.\mathtt{Use_1}$$

There is now a similar argument. The process $\overline{\mathtt{little}}.\mathtt{Use_1}$ can only contribute to an action in $[-K^+]$ if it is part of a communication. Therefore, one needs to show the following pair.

$$\mathtt{Ven_1} \models [-K^+](\langle\mathtt{collect_1}\rangle\mathtt{tt}/\overline{\mathtt{little}}.\mathtt{Use_1})$$
$$\mathtt{Ven_1} \models [\mathtt{little}](\langle\mathtt{collect_1}\rangle\mathtt{tt}/\mathtt{Use_1})$$

The first is derivable using Proposition 1, and by the same Proposition the second is equivalent to the following.

$$\mathtt{collect_1.Ven} \models (\langle\mathtt{collect_1}\rangle\mathtt{tt})/\mathtt{Use_1}$$

This clearly holds because $\mathtt{collect_1.Ven}$ is able to perform $\mathtt{collect_1}$, and $\mathtt{tt}/F$ for any process F is just the formula $\mathtt{tt}$.

Exercises

1. Using Proposition 1 and the semantic clauses for boolean connectives, show the following.
   - **a.** $\mathtt{Cl} \models [\mathtt{tick},\mathtt{tock}](\langle\mathtt{tick}\rangle\mathtt{tt} \wedge [\mathtt{tock}]\mathtt{ff})$
   - **b.** $\mathtt{Ven} \models [2\mathrm{p}, 1\mathrm{p}]\langle\mathtt{big},\mathtt{little}\rangle\mathtt{tt}$
   - **c.** $\mathtt{Ct_0} \models [\mathtt{up}][\mathtt{down},\mathtt{up}]([\mathtt{down}]\mathtt{ff} \vee [\mathtt{round}]\mathtt{ff})$
2. Let $K = \{a, b, c\}$. What are the following formulas?
   - **a.** $(\langle\tau\rangle[a]\langle b\rangle\mathtt{tt} \wedge [-][-]\langle c\rangle\mathtt{tt})\backslash K$
   - **b.** $([a, \overline{b}, c, d]\langle d\rangle[\overline{c}, d]\mathtt{ff} \vee [-J][-K]\mathtt{ff})\backslash K$
3. Define operators $[f]$ and $\backslash\backslash J$ on modal formulas so that the following hold.
   - **a.** $E[f] \models \Phi$ iff $E \models \Phi[f]$
   - **b.** $E\backslash\backslash J \models \Phi$ iff $E \models \Phi\backslash\backslash J$
4. Define the slicing operator $/F$ on modal formulas. Use your definition to prove the following without appealing to transitional behaviour.
   - **a.** $\mathtt{Crossing} \models [\mathtt{train}](\langle\tau\rangle\mathtt{tt} \wedge \langle\mathtt{car}\rangle[\tau][\tau]\mathtt{ff})$
   - **b.** $\mathtt{Crossing} \models [\mathtt{car}][\mathtt{train}][\tau](\langle\overline{\mathtt{tcross}}\rangle\mathtt{tt} \vee \langle\overline{\mathtt{ccross}}\rangle\mathtt{tt})$
   - **c.** $\mathtt{Crossing} \not\models [\mathtt{car}][\mathtt{train}][\tau](\langle\overline{\mathtt{tcross}}\rangle\mathtt{tt} \wedge \langle\overline{\mathtt{ccross}}\rangle\mathtt{tt})$

2.4 Observable modal logic

Process activity is delineated by the two kinds of transition relation distinguished by the thickness of their arrows, $\longrightarrow$ and $\Longrightarrow$. The latter captures the operation of observable transitions because $\stackrel{a}{\Longrightarrow}$ permits silent activity before and after a happens. Thc relation $\stackrel{a}{\Longrightarrow}$ (see Section 1.3) was defined in terms of $\stackrel{a}{\longrightarrow}$ and the relation $\stackrel{\varepsilon}{\Longrightarrow}$ indicating zero or more silent actions.

The modal logic M does not express observable capabilities of processes because silent actions are not accorded a special status. To overcome this, it suffices to introduce new modalities $[\![\ ]\!]$ and $\langle\!\langle\ \rangle\!\rangle$ as follows.

$$E \models [\![\ ]\!]\Phi \quad \text{iff} \quad \forall F \in \{E' : E \stackrel{\varepsilon}{\Longrightarrow} E'\}.\ F \models \Phi$$
$$E \models \langle\!\langle\ \rangle\!\rangle\Phi \quad \text{iff} \quad \exists F \in \{E' : E \stackrel{\varepsilon}{\Longrightarrow} E'\}.\ F \models \Phi$$

A process has the property $[\![\ ]\!]\,\Phi$ provided that it satisfies Φ and, after evolving through any amount of silent activity, Φ remains true. To satisfy $\langle\!\langle\ \rangle\!\rangle\Phi$, a process has to be able to evolve in zero or more τ transitions to a process realizing Φ.

Neither $[\![\ ]\!]$ nor $\langle\!\langle\ \rangle\!\rangle$ is definable within the modal logic M. A technique for showing non-definability employs equivalence of formulas: two formulas Φ and Ψ are equivalent if, for every process E, $E \models \Phi$ iff $E \models \Psi$. For instance, $\langle\texttt{tick}, \texttt{tock}\rangle\Phi$ is equivalent to $\langle\texttt{tick}\rangle\Phi \vee \langle\texttt{tock}\rangle\Phi$, for any Φ. Two formulas are *not* equivalent if there is a process that realises one, but not the other, formula. If $[\![\ ]\!]$ is definable in M, then for any formula Φ in M there is also a formula in M equivalent to $[\![\ ]\!]\,\Phi$ (and similarly for definability of $\langle\!\langle\ \rangle\!\rangle$). Non-definability of $[\![\ ]\!]$ is established if there is a Φ in M and $[\![\ ]\!]\,\Phi$ is not equivalent to *any* formula of M. A simple choice of Φ, namely $[a]\texttt{ff}$, suffices. A process realises $[\![\ ]\!]\,[a]\texttt{ff}$ if it is unable to perform a after any amount of silent activity. We show that $[\![\ ]\!]\,[a]\texttt{ff}$ is not equivalent to any formula of M. For this purpose, consider the two families of similar processes $\{\texttt{Div}_i : i \in \mathbb{N}\}$ and $\{\texttt{Div}_i^a : i \in \mathbb{N}\}$

$$\begin{array}{ll} \texttt{Div}_0 \stackrel{\text{def}}{=} \tau.0 & \texttt{Div}_{i+1} \stackrel{\text{def}}{=} \tau.\texttt{Div}_i \\ \texttt{Div}_0^a \stackrel{\text{def}}{=} a.0 & \texttt{Div}_{i+1}^a \stackrel{\text{def}}{=} \tau.\texttt{Div}_i^a \end{array}$$

whose transition graphs are as follows.

$$\ldots \stackrel{\tau}{\longrightarrow} \texttt{Div}_{i+1} \stackrel{\tau}{\longrightarrow} \texttt{Div}_i \ldots \stackrel{\tau}{\longrightarrow} \texttt{Div}_1 \stackrel{\tau}{\longrightarrow} \texttt{Div}_0 \stackrel{\tau}{\longrightarrow} 0$$

$$\ldots \stackrel{\tau}{\longrightarrow} \texttt{Div}_{i+1}^a \stackrel{\tau}{\longrightarrow} \texttt{Div}_i^a \ldots \stackrel{\tau}{\longrightarrow} \texttt{Div}_1^a \stackrel{\tau}{\longrightarrow} \texttt{Div}_0^a \stackrel{a}{\longrightarrow} 0$$

$\texttt{Div}_n \models [\![\ ]\!]\,[a]\texttt{ff}$ for each n. On the other hand, $\texttt{Div}_n^a \not\models [\![\ ]\!]\,[a]\texttt{ff}$ for each n because of the transition $\texttt{Div}_n^a \stackrel{\varepsilon}{\Longrightarrow} \texttt{Div}_0^a$. For each formula Ψ of M, there is a $k \geq 0$ with the feature that $\texttt{Div}_k \models \Psi$ iff $\texttt{Div}_k^a \models \Psi$, and therefore $[\![\ ]\!]\,[a]\texttt{ff}$ is not equivalent to any M formula. The crucial step here in a strengthened form is demonstrated in Proposition 1, below. Recall from Section 2.2 that the size of $|\Psi|$ is the number of occurrences of $\texttt{tt}$, $\texttt{ff}$, $\wedge$, $\vee$, $[K]$ and $\langle K\rangle$ within it.

Proposition 1 *If $\Psi \in$ M and $|\Psi| = k$, then for all $m \geq k$, $\texttt{Div}_m \models \Psi$ iff $\texttt{Div}_m^a \models \Psi$.*

Proof. By induction on k. The base case is $k = 1$, so Ψ is $\mathtt{tt}$ or $\mathtt{ff}$, and clearly the property holds. For the induction step, assume the result for all $k \leq n$. Suppose $|\Psi| = n + 1$. Four cases need to be dealt with. If Ψ is $\Psi_1 \wedge \Psi_2$ or $\Psi_1 \vee \Psi_2$, then as each component Ψ_i, $i \in \{1, 2\}$, has size less than $n + 1$ the induction hypothesis applies to it. It follows that $\mathtt{Div}_m \models \Psi_i$ iff $\mathtt{Div}^a_m \models \Psi_i$ for all $m \geq n + 1$. Now the result follows. Otherwise, Ψ is $[K]\Psi_1$ or $\langle K\rangle\Psi_1$. We just consider the first of these cases and leave the second as an exercise for the reader. If $\tau \notin K$, then for all $m \geq 1$, $\mathtt{Div}_m \models \Psi$ and $\mathtt{Div}^a_m \models \Psi$. So, assume $\tau \in K$. As $|\Psi_1| = n$, by the induction hypothesis for all $m \geq n$, $\mathtt{Div}_m \models \Psi_1$ iff $\mathtt{Div}^a_m \models \Psi_1$. And therefore for all $m \geq n$, $\mathtt{Div}_{m+1} \models [K]\Psi_1$ iff $\mathtt{Div}^a_{m+1} \models [K]\Psi_1$. □

Using the new modal operators, supplementary modalities $[\![K]\!]$ and $\langle\!\langle K\rangle\!\rangle$ are definable as follows, when K is a subset of *observable* actions O.

$$[\![K]\!]\,\Phi \stackrel{\text{def}}{=} [\![\;]\!][K][\![\;]\!]\,\Phi \qquad\qquad \langle\!\langle K\rangle\!\rangle\Phi \stackrel{\text{def}}{=} \langle\!\langle\;\rangle\!\rangle\langle K\rangle\langle\!\langle\;\rangle\!\rangle\Phi$$

Their meanings appeal to observable transition relations $\stackrel{a}{\Longrightarrow}$ in the same way that the meanings of $[K]$ and $\langle K\rangle$ appeal to the relations $\stackrel{a}{\longrightarrow}$. Process E has the property $[\![K]\!]\,\Phi$

$$\begin{array}{ll}
\text{iff} & E \models [\![\;]\!][K][\![\;]\!] \\
\text{iff} & \forall F \in \{E' : E \stackrel{\varepsilon}{\Longrightarrow} E'\}.\ F \models [K][\![\;]\!]\,\Phi \\
\text{iff} & \forall F \in \{E' : E \stackrel{\varepsilon}{\Longrightarrow} E_1 \stackrel{a}{\longrightarrow} E' \text{ and } a \in K\}.\ F \models [\![\;]\!]\,\Phi \\
\text{iff} & \forall F \in \{E' : E \stackrel{\varepsilon}{\Longrightarrow} E_1 \stackrel{a}{\longrightarrow} E_2 \stackrel{\varepsilon}{\Longrightarrow} E' \text{ and } a \in K\}.\ F \models \Phi \\
\text{iff} & \forall F \in \{E' : E \stackrel{a}{\Longrightarrow} E' \text{ and } a \in K\}.\ F \models \Phi,
\end{array}$$

and $E \models \langle\!\langle K\rangle\!\rangle\Phi$ iff $\exists F \in \{E' : E \stackrel{a}{\Longrightarrow} E' \text{ and } a \in K\}.\ F \models \Phi$. As with the modalities of Section 2.1, we write $[\![a_1, \ldots, a_n]\!]$ and $\langle\!\langle a_1, \ldots, a_n\rangle\!\rangle$ instead of $[\![\{a_1, \ldots, a_n\}]\!]$ and $\langle\!\langle\{a_1, \ldots, a_n\}\rangle\!\rangle$.

The simple modal formula $\langle\!\langle\mathtt{tick}\rangle\!\rangle\mathtt{tt}$ expresses the observable ability to carry out the action $\mathtt{tick}$, whereas $[\![\mathtt{tick}]\!]\,\mathtt{ff}$ expresses an inability to tick after any amount of internal activity. Both clocks $\mathtt{Cl}$ and $\mathtt{Cl}_5$ from Section 1.1, where $\mathtt{Cl}_5 \stackrel{\text{def}}{=} \mathtt{tick}.\mathtt{Cl}_5 + \tau.0$, have the property $\langle\!\langle\mathtt{tick}\rangle\!\rangle\mathtt{tt}$, but the clock $\mathtt{Cl}_5$ may at any time silently stop ticking, and therefore also has the property $\langle\!\langle\mathtt{tick}\rangle\!\rangle\,[\![\mathtt{tick}]\!]\,\mathtt{ff}$.

Example 1 The crossing of Section 1.2 has the property that, after a car and a train approach, one of them may cross.

$$[\![\mathtt{car}]\!]\,[\![\mathtt{train}]\!]\,(\langle\!\langle\overline{\mathtt{tcross}}\rangle\!\rangle\mathtt{tt} \vee \langle\!\langle\overline{\mathtt{ccross}}\rangle\!\rangle\mathtt{tt})$$

In the following the processes, E_i are from Figure 1.12.

$$\begin{aligned}
& \texttt{Crossing} \models [\![\texttt{car}]\!]\,[\![\texttt{train}]\!]\,(\langle\!\langle\overline{\texttt{tcross}}\rangle\!\rangle\texttt{tt} \vee \langle\!\langle\overline{\texttt{ccross}}\rangle\!\rangle\texttt{tt}) \\
\text{iff}\quad & E_1 \models [\![\texttt{train}]\!]\,(\langle\!\langle\overline{\texttt{tcross}}\rangle\!\rangle\texttt{tt} \vee \langle\!\langle\overline{\texttt{ccross}}\rangle\!\rangle\texttt{tt}) \ \text{ and} \\
& E_4 \models [\![\texttt{train}]\!]\,(\langle\!\langle\overline{\texttt{tcross}}\rangle\!\rangle\texttt{tt} \vee \langle\!\langle\overline{\texttt{ccross}}\rangle\!\rangle\texttt{tt}) \\
\text{iff}\quad & E_3 \models \langle\!\langle\overline{\texttt{tcross}}\rangle\!\rangle\texttt{tt} \vee \langle\!\langle\overline{\texttt{ccross}}\rangle\!\rangle\texttt{tt} \ \text{ and} \\
& E_6 \models \langle\!\langle\overline{\texttt{tcross}}\rangle\!\rangle\texttt{tt} \vee \langle\!\langle\overline{\texttt{ccross}}\rangle\!\rangle\texttt{tt} \ \text{ and} \\
& E_7 \models \langle\!\langle\overline{\texttt{tcross}}\rangle\!\rangle\texttt{tt} \vee \langle\!\langle\overline{\texttt{ccross}}\rangle\!\rangle\texttt{tt}
\end{aligned}$$

Both E_3 and E_7 have observable $\overline{\texttt{tcross}}$ transitions, and E_6 has an observable $\overline{\texttt{ccross}}$ transition.

The set O, introduced in Section 1.3, is a universal set of observable actions (which does not contain τ). We assume the following abbreviations.

$$\begin{aligned}
[\![-K]\!]\,\Phi & \stackrel{\text{def}}{=} [\![\mathsf{O} - K]\!] \\
\langle\!\langle -K\rangle\!\rangle\Phi & \stackrel{\text{def}}{=} \langle\!\langle \mathsf{O} - K\rangle\!\rangle
\end{aligned}$$

Therefore, $[\![-]\!]$ and $\langle\!\langle-\rangle\!\rangle$ are abbreviations of $[\![\mathsf{O}]\!]$ and $\langle\!\langle\mathsf{O}\rangle\!\rangle$. The modal formula $[\![-]\!]\,\texttt{ff}$ hence expresses an inability to carry out an observable action, so the process $\texttt{Div} \stackrel{\text{def}}{=} \tau.\texttt{Div}$ realises it.

Modal formulas can be used to express notions that are basic to the theory of CSP [31]. A process can carry out the observable trace $a_1 \ldots a_n$ provided it has the property $\langle\!\langle a_1\rangle\!\rangle \ldots \langle\!\langle a_n\rangle\!\rangle\texttt{tt}$. A process is stable if it has no τ-transitions (and, therefore, if it has the property $[\tau]\texttt{ff}$). The formula $[\![K]\!]\,\texttt{ff}$ expresses that the observable set of actions K is a "refusal," since a realizing process is unable to perform observable actions belonging to K. The pair $(a_1 \ldots a_n, K)$ is an observable failure for a process provided it satisfies $\langle\!\langle a_1\rangle\!\rangle \ldots \langle\!\langle a_n\rangle\!\rangle([\tau]\texttt{ff} \wedge [\![K]\!]\,\texttt{ff})$: a process realises this formula if it can carry out the observable trace $a_1 \ldots a_n$ and become a stable process that is unable to carry out observable actions belonging to K.

Example 2 The processes Cop and Protocol have the same observable failures, as given by the following sequence of formulas.

$$\begin{aligned}
& \langle\!\langle\ \rangle\!\rangle([\tau]\texttt{ff} \wedge [\![-\{\texttt{in}(m)\ :\ m \in D\}]\!]\,\texttt{ff}) \\
& \langle\!\langle\texttt{in}(m)\rangle\!\rangle([\tau]\texttt{ff} \wedge [\![-\overline{\texttt{out}}(m)]\!]\,\texttt{ff}) \\
& \langle\!\langle\texttt{in}(m)\rangle\!\rangle\langle\!\langle\overline{\texttt{out}}(m)\rangle\!\rangle([\tau]\texttt{ff} \wedge [\![-\{\texttt{in}(n)\ :\ n \in D\}]\!]\,\texttt{ff}) \\
& \ \vdots \qquad\quad \vdots
\end{aligned}$$

For instance, Protocol satisfies the second formula because of the following transition.

$$\texttt{Protocol} \stackrel{\texttt{in}(m)}{\Longrightarrow} (\texttt{Send1}(m) \mid \texttt{Medium} \mid \overline{\texttt{out}}(m).\overline{\texttt{ok}}.\texttt{Receiver})\backslash J,$$

where $J = \{\texttt{sm}, \texttt{ms}, \texttt{mr}, \texttt{ok}\}$.

There are two transition graphs asssociated with any process, as described in Section 1.3, one built from thin transitions $\xrightarrow{a}$ and $\xrightarrow{\tau}$, and the other from the observable transitions $\xRightarrow{a}$ and $\xRightarrow{\varepsilon}$. The modal logic M is associated with the first kind of graph. For the second kind, we introduce observable modal logic M^o, whose formulas are defined inductively below.

$$\Phi ::= \mathtt{tt} \mid \mathtt{ff} \mid \Phi_1 \wedge \Phi_2 \mid \Phi_1 \vee \Phi_2 \mid [\![K]\!]\,\Phi \mid [\![\ ]\!]\,\Phi \mid \langle\!\langle K\rangle\!\rangle\Phi \mid \langle\!\langle\ \rangle\!\rangle\Phi$$

K ranges over subsets of O. The logic M^o is closed under complement, as the reader can ascertain (for instance $([\![K]\!]\,\Phi)^c$ is $\langle\!\langle K\rangle\!\rangle\Phi^c$ and $(\langle\!\langle\ \rangle\!\rangle\Phi)^c$ is $[\![\ ]\!]\,\Phi^c$).

Exercises

1. Show that both $[\![\ ]\!]$ and $\langle\!\langle\ \rangle\!\rangle$ are definable within infinitary modal logic M_∞ of Section 2.2.
2. Show that $\langle\!\langle\ \rangle\!\rangle$ is not definable in M by employing a similar argument to that used in Proposition 1, but with respect to the dual formula $\langle\!\langle\ \rangle\!\rangle\langle a\rangle\mathtt{tt}$.
3. The modal depth of a formula Ψ, written $\mathrm{md}(\Psi)$, is defined inductively as follows.

$$\begin{array}{lclcl} \mathrm{md}(\mathtt{tt}) & = & 0 & = & \mathrm{md}(\mathtt{ff}) \\ \mathrm{md}(\Phi_1 \wedge \Phi_2) & = & \max\{\mathrm{md}(\Phi_1), \mathrm{md}(\Phi_2)\} & = & \mathrm{md}(\Phi_1 \vee \Phi_2) \\ \mathrm{md}([K]\Phi) & = & 1 + \mathrm{md}(\Phi) & = & \mathrm{md}(\langle K\rangle\Phi) \end{array}$$

   Show that Proposition 1 remains true if $|\Psi|$ is replaced with $\mathrm{md}(\Psi)$.
4. Let M^- represent the family of modal formulas of M that do not contain occurrences of box modalities, $[J]$ for any J. Show that, for any non-empty J, the modality $[J]$ is not definable in M^-.
5. Prove the following pair.
   a. $\mathtt{Crossing} \not\models [\![\mathtt{car}]\!]\,[\![\mathtt{train}]\!]\,(\langle\!\langle\overline{\mathtt{tcross}}\rangle\!\rangle\mathtt{tt} \wedge \langle\!\langle\overline{\mathtt{ccross}}\rangle\!\rangle\mathtt{tt})$
   b. $\mathtt{Protocol} \models [\![\mathtt{in}(m)]\!]\,(\langle\!\langle\overline{\mathtt{out}}(m)\rangle\!\rangle\mathtt{tt} \wedge [\![\{\mathtt{in}(m) : m \in D\}]\!]\,\mathtt{ff})$
6. Consider the following three vending machines.

$$\begin{array}{lcl} \mathtt{Ven}_1 & \stackrel{\mathrm{def}}{=} & \mathtt{1p.1p.(tea.Ven_1 + coffee.Ven_1)} \\ \mathtt{Ven}_2 & \stackrel{\mathrm{def}}{=} & \mathtt{1p.(1p.tea.Ven_2 + 1p.coffee.Ven_2)} \\ \mathtt{Ven}_3 & \stackrel{\mathrm{def}}{=} & \mathtt{1p.1p.tea.Ven_3 + 1p.1p.coffee.Ven_3} \end{array}$$

   Show that $\mathtt{Ven}_2$ and $\mathtt{Ven}_3$ have the same set of observable failures, but that $\mathtt{Ven}_1$ and $\mathtt{Ven}_2$ have different observable failures.
7. Show that $(\mathtt{C} \mid \mathtt{U})\backslash\{\mathtt{in}, \mathtt{ok}\}$ and $\mathtt{Ucop}$ from Section 1.3 have the same M^o modal properties, but not the same M properties.
8. Show that for all $\Phi \in \mathrm{M}^o$, $\mathtt{Cop} \models \Phi$ iff $\mathtt{Protocol} \models \Phi$.
9. An M^o formula is realisable if there is a process that satisfies it. Which of the following formulas are realisable?

a. $\langle\!\langle \mathtt{car} \rangle\!\rangle \langle\!\langle \mathtt{train} \rangle\!\rangle (\langle\!\langle \overline{\mathtt{tcross}} \rangle\!\rangle \mathtt{tt} \vee \langle\!\langle \overline{\mathtt{ccross}} \rangle\!\rangle \mathtt{tt})$

b. $[\![\ ]\!]\,\mathtt{ff}$

c. $\langle\!\langle \mathtt{car} \rangle\!\rangle \langle\!\langle \mathtt{train} \rangle\!\rangle (\langle\!\langle \overline{\mathtt{tcross}} \rangle\!\rangle \mathtt{tt} \wedge \langle\!\langle \overline{\mathtt{ccross}} \rangle\!\rangle \mathtt{tt})$

d. $\langle\!\langle \mathtt{car} \rangle\!\rangle \langle\!\langle \mathtt{train} \rangle\!\rangle (\langle\!\langle\ \rangle\!\rangle \langle\!\langle \overline{\mathtt{ccross}} \rangle\!\rangle \mathtt{tt} \wedge \langle\!\langle\ \rangle\!\rangle\, [\![\overline{\mathtt{ccross}}]\!]\, \mathtt{ff})$

e. $\langle\!\langle \mathtt{car} \rangle\!\rangle \langle\!\langle \mathtt{train} \rangle\!\rangle (\langle\!\langle\ \rangle\!\rangle \langle\!\langle \overline{\mathtt{ccross}} \rangle\!\rangle \mathtt{tt} \wedge [\![\overline{\mathtt{ccross}}]\!]\, \mathtt{ff})$

2.5 Observable necessity and divergence

Within the modal logics M and M^o, capabilities and observable capabilities of processes are expressible. M also permits expression of immediate necessity (or inevitability). The formula $\langle - \rangle \mathtt{tt} \wedge [-a]\mathtt{ff}$ expresses that a must be the very next action because $\langle - \rangle \mathtt{tt}$ asserts that some action is possible, whereas $[-a]\mathtt{ff}$ states that only a is a possible next action. In this section we examine how to express immediate observable necessity.

A first attempt at expressing in M^o the property that a must be the next observable action is $\langle\!\langle - \rangle\!\rangle \mathtt{tt} \wedge [\![-a]\!]\, \mathtt{ff}$, which states that some observable action is possible and that all observable actions except a are initially impossible. However, it leaves open the possibility that a may become excluded through silent activity. Both clocks $\mathtt{Cl}$ and $\mathtt{Cl}_5$ satisfy $\langle\!\langle - \rangle\!\rangle \mathtt{tt} \wedge [\![-\mathtt{tick}]\!]\, \mathtt{ff}$, as mentioned in the previous section. $\mathtt{Cl}_5$ is able to carry out an observable $\mathtt{tick}$ transition, $\mathtt{Cl}_5 \stackrel{\mathtt{tick}}{\Longrightarrow} \mathtt{Cl}_5$, and also is unable to perform any other observable action. However, this clock may also silently break down, $\mathtt{Cl}_5 \stackrel{\varepsilon}{\Longrightarrow} \mathtt{0}$, and become unable to tick. This shortcoming can be surmounted by strengthening the initial conjunct $\langle\!\langle - \rangle\!\rangle \mathtt{tt}$ to $[\![\ ]\!] \langle\!\langle - \rangle\!\rangle \mathtt{tt}$, requiring that an observable action be possible after any amount of silent activity. $\mathtt{Cl}_5 \not\models [\![\ ]\!] \langle\!\langle - \rangle\!\rangle \mathtt{tt}$ because of the silent transition $\mathtt{Cl}_5 \stackrel{\varepsilon}{\Longrightarrow} \mathtt{0}$, whereas $\mathtt{Cl}$ has this property.

A second attempt at expressing necessity of a is $[\![\ ]\!] \langle\!\langle - \rangle\!\rangle \mathtt{tt} \wedge [\![-a]\!]\, \mathtt{ff}$. Certainly, this formula expresses that the initial observable action must be a. However, it does not guarantee that a first observable action will happen. $\mathtt{Cl}_6$ is another clock, $\mathtt{Cl}_6 \stackrel{\mathrm{def}}{=} \mathtt{tick}.\mathtt{Cl}_6 + \tau.\mathtt{Cl}_6$, which satisfies $[\![\ ]\!] \langle\!\langle - \rangle\!\rangle \mathtt{tt} \wedge [\![-\mathtt{tick}]\!]\, \mathtt{ff}$. It realizes both conjuncts because its only observable transitions are $\mathtt{Cl}_6 \stackrel{\mathtt{tick}}{\Longrightarrow} \mathtt{Cl}_6$ and $\mathtt{Cl}_6 \stackrel{\varepsilon}{\Longrightarrow} \mathtt{Cl}_6$. Interpreting this formula as an inevitability that $\mathtt{tick}$ happens next fails to take into account the possibility that $\mathtt{Cl}_6$ avoids ticking by perpetually engaging in silent activity.

Example 1 $\mathtt{Cop}$ and $\mathtt{Protocol}$ both have the property that, after an input, an output must happen next, that is, for any m

$$[\![\mathtt{in}(m)]\!] ([\![\ ]\!] \langle\!\langle - \rangle\!\rangle \mathtt{tt} \wedge [\![-\overline{\mathtt{out}}(m)]\!]\, \mathtt{ff}).$$

In the case of `Protocol`, the message m may also be continually lost during transmission, and therefore may never be transmitted. This is not possible for `Cop`.

A process *diverges* if it is able to perform internal actions forever. $\texttt{Cl}_6$ diverges because of the following endless sequence of τ-transitions.

$$\texttt{Cl}_6 \xrightarrow{\tau} \texttt{Cl}_6 \xrightarrow{\tau} \ldots \xrightarrow{\tau} \texttt{Cl}_6 \xrightarrow{\tau} \ldots$$

In contrast, $\texttt{Cl}_5$ does not diverge, and so is said to *converge*. Following Hennessy [26], let $E \uparrow$ abbreviate that E diverges, and $E \downarrow$ abbreviate that E converges. Neither convergence nor divergence is definable in the modal logics M and M^o. There is no formula Φ in these logics such that, for every process E, $E \models \Phi$ iff $E \uparrow$. Consequently, we introduce another pair of modalities $[\![\downarrow]\!]$ and $\langle\!\langle\uparrow\rangle\!\rangle$, analogous to $[\![\ ]\!]$ and $\langle\!\langle\ \rangle\!\rangle$, except that they contain information about divergence and convergence.

$$\begin{array}{lll} E \models [\![\downarrow]\!]\,\Phi & \text{iff} & E\downarrow \text{ and } \forall F \in \{E' : E \stackrel{\varepsilon}{\Longrightarrow} E'\}.\ F \models \Phi \\ E \models \langle\!\langle\uparrow\rangle\!\rangle\Phi & \text{iff} & E\uparrow \text{ or } \exists F \in \{E' : E \stackrel{\varepsilon}{\Longrightarrow} E'\}.\ F \models \Phi \end{array}$$

A process satisfies $[\![\downarrow]\!]\,\Phi$ if it converges and realises $[\![\ ]\!]\,\Phi$. Dually, a process realises $\langle\!\langle\uparrow\rangle\!\rangle\Phi$ if it diverges or satisfies $\langle\!\langle\ \rangle\!\rangle\Phi$. Divergence and convergence are expressible by means of these new modalities: $[\![\downarrow]\!]\,\texttt{tt}$ expresses convergence and its dual $\langle\!\langle\uparrow\rangle\!\rangle\texttt{ff}$ expresses divergence.

Let $\mathrm{M}^{o\downarrow}$ be observable modal logic together with the two new modalities:

$$\begin{array}{rcl} \Phi & ::= & \texttt{tt} \mid \texttt{ff} \mid \Phi_1 \wedge \Phi_2 \mid \Phi_1 \vee \Phi_2 \mid [\![K]\!]\,\Phi \mid [\![\ ]\!]\,\Phi \mid [\![\downarrow]\!]\,\Phi \mid \\ & & \langle\!\langle K\rangle\!\rangle\Phi \mid \langle\!\langle\ \rangle\!\rangle\Phi \mid \langle\!\langle\uparrow\rangle\!\rangle\Phi \end{array}$$

Within $\mathrm{M}^{o\downarrow}$, the strong observable inevitability that a must (and will) happen next is expressible as

$$[\![\downarrow]\!]\,\langle\!\langle-\rangle\!\rangle\texttt{tt} \wedge [\![-\texttt{a}]\!]\,\texttt{ff}.$$

The initial conjunct precludes the possibility that τ could occur forever. Consequently, $\texttt{Cl}_6 \not\models [\![\downarrow]\!]\,\langle\!\langle-\rangle\!\rangle\texttt{tt} \wedge [\![-\texttt{tick}]\!]\,\texttt{ff}$.

Example 2 The difference between `Cop` and `Protocol` as described in Example 1 is expressed as $[\![\texttt{in}(m)]\!]\,([\![\downarrow]\!]\,\langle\!\langle-\rangle\!\rangle\texttt{tt} \wedge \big[\!\big[-\overline{\texttt{out}}(m)\big]\!\big]\,\texttt{ff})$ for any m. `Protocol` fails to have this property because the message m may be continually lost during transmission, and therefore may never be output.

Ancillary modalities can be defined in $M^{o\downarrow}$ as follows.

$$\begin{array}{lcl} [\![\downarrow K]\!]\,\Phi & \stackrel{\text{def}}{=} & [\![\downarrow]\!]\,[\![K]\!]\,\Phi \\ [\![K\downarrow]\!]\,\Phi & \stackrel{\text{def}}{=} & [\![K]\!]\,[\![\downarrow]\!]\,\Phi \\ [\![\downarrow K\downarrow]\!]\,\Phi & \stackrel{\text{def}}{=} & [\![\downarrow]\!]\,[\![K]\!]\,[\![\downarrow]\!]\,\Phi \end{array}$$

Features of processes dealing with divergence, appealed to in definitions of behavioural refinement [26], can also be expressed as modal formulas in this extended modal logic. For instance, that E cannot diverge throughout the observable trace $a_1 \ldots a_n$ is captured as $E \models [\![\downarrow a_1 \downarrow]\!] \ldots [\![\downarrow a_n \downarrow]\!]\,\mathtt{tt}$.

Exercises

1. Show that the modalities $[\![\downarrow]\!]$ and $\langle\!\langle\uparrow\rangle\!\rangle$ are not definable in M^o.
2. Show that Cop realises the property

   $[\![\mathtt{in}(m)]\!]\,([\![\downarrow]\!]\,\langle\!\langle-\rangle\!\rangle\mathtt{tt} \wedge [\![-\overline{\mathtt{out}}(m)]\!]\,\mathtt{ff})$

   and Protocol fails to have this property.
3. Prove that, $\mathrm{M}^{o\downarrow}$ is closed under complement.
4. Prove that, for all Φ in M^o, $0 \models \Phi$ iff $\mathtt{Div} \models \Phi$.
5. Prove that for all Φ in $\mathrm{M}^{o\downarrow}$, $\mathtt{Ct}_0 \models \Phi$ iff $\mathtt{Count} \models \Phi$.
6. Which of the following $\mathrm{M}^{o\downarrow}$ formulas are realisable?
   a. $[\![K]\!]\,([\![\downarrow]\!]\,\langle\!\langle\mathtt{tick}\rangle\!\rangle\mathtt{tt} \wedge \langle\!\langle\uparrow\rangle\!\rangle\langle\!\langle\mathtt{tick}\rangle\!\rangle\mathtt{tt})$
   b. $[\![K\downarrow]\!]\,\langle\!\langle\uparrow\rangle\!\rangle\,[\![\mathtt{tick}]\!]\,\mathtt{ff}$
   c. $[\![K]\!]\,(\langle\!\langle\uparrow\rangle\!\rangle\langle\!\langle\mathtt{tick}\rangle\!\rangle\mathtt{tt} \wedge [\![\;]\!]\,[\![\mathtt{tick}]\!]\,\mathtt{ff})$
   d. $[\![K\downarrow]\!]\,(\langle\!\langle\uparrow\rangle\!\rangle\langle\!\langle\mathtt{tick}\rangle\!\rangle\mathtt{tt} \wedge [\![\;]\!]\,[\![\mathtt{tick}]\!]\,\mathtt{ff})$

3

Bisimulations

Example processes were defined in Chapter 1, and in Chapter 2 modal logics were introduced for expressing their capabilities. An important issue arises when two processes may be deemed to have the same behaviour. Such an abstraction can be presented by defining an appropriate equivalence relation between processes. In this chapter, we focus on equivalences for CCS processes defined in terms of bisimulation relations. However, we present them using games that provide a powerful metaphor for understanding interaction. There is also an intimate relation between modal properties and these equivalences.

3.1 Process equivalences

Process expressions are intended to be used for describing interacting systems. So far, the discussion has omitted criteria applicable to when two expressions may be said for all intents and purposes to describe the same system. Alternatively, we can consider grounds for differentiating process descriptions. Undoubtedly, the clock `Cl` and the vending machine `Ven` of Section 1.1 are different. They are intended as models of distinct kinds of objects. At all levels of description they differ: in their algebraic expressions, their action names, their flow graphs and their transition graphs. A concrete manifestation of their difference is their initial capabilities. The clock Cl can perform the (observable) action tick, whereas Ven can not.

Syntactic differences alone should not be sufficient grounds for distinguishing processes. It is important to allow the possibility that two process descriptions may be equivalent, even though they may differ markedly in their level of detail. An example is that of two descriptions of a counter $\mathtt{Ct}_0$ of Figure 1.4 and Count of Section 1.6. An account of process equivalence has practical significance when one views process expressions both as specifications and as descriptions of implementations. $\mathtt{Ct}_0$ is a specification (even the requirement specification) of a counter, whereas the finite description Count with its very different structure can be seen as a description of a possible implementation. Similarly, the buffer Cop can be seen as a specification of the process Protocol of Section 1.2. In this context, an account of process equivalence could tell us when an implementation meets its specification[1].

Example 1 A stark description of the slot machine $\mathtt{SM}_n$ of Figure 1.15 is the process $\mathtt{SM}'_n$

$$\mathtt{SM}'_i \stackrel{\mathrm{def}}{=} \mathtt{slot}.(\tau.\overline{\mathtt{loss}}.\mathtt{SM}'_{i+1} + \sum\{\tau.\overline{\mathtt{win}}(y).\mathtt{SM}'_{(i+1)-y} \ : \ 1 \leq y \leq i+1\}),$$

which carries no assumptions as to how the slot machine is to be built from separate but interacting concurrent components.

The two counters $\mathtt{Ct}_0$ and Count have the same flow graph. Not only do they have the same initial observable capabilities, but this feature is also preserved as observable actions are performed. There is a similarity between their observable transition graphs, a resemblance not immediately easy to define. A much simpler case is that of the two clocks, $\mathtt{Cl} \stackrel{\mathrm{def}}{=} \mathtt{tick.Cl}$ and $\mathtt{Cl}_2 \stackrel{\mathrm{def}}{=} \mathtt{tick.tick.Cl}_2$ pictured in Figure 3.1. Although they have different transition graphs, whatever transitions one of these clocks makes can be matched by the other, and the resulting processes also retain this property. An alternative basis for suggesting that these two clocks are equivalent starts with the observation that Cl and tick.Cl should count as equivalent expressions because Cl is defined as tick.Cl. An important principle

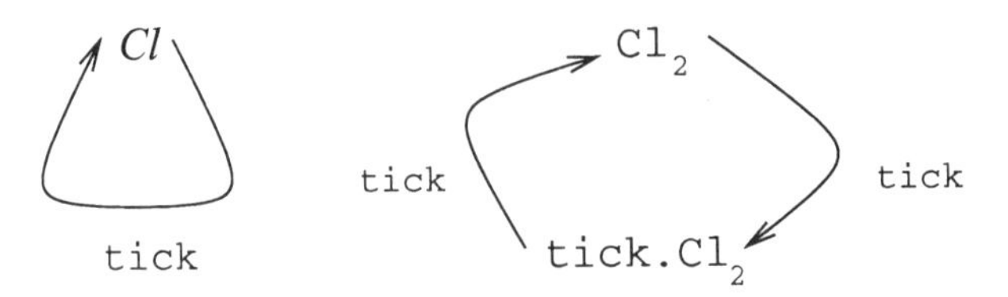

FIGURE 3.1. Two clocks

[1] Similar comments could be made about refinement, where we would expect an ordering on processes.

is that, if two expressions are equivalent, then replacing one with the other in some other expression should preserve equivalence. Replacing `Cl` with `tick.Cl` in the expression E should result in a process expression equivalent to E. In particular, if E is `tick.Cl` then `tick.tick.Cl` and `tick.Cl` should count as equivalent. Because particular names of processes are unimportant, this should imply that `Cl` and $\mathtt{Cl}_2$ are also equivalent.

The extensionality principle here is that an equivalence should also be a congruence. That is, the equivalence should be preserved by the various process combinators. For instance, if E and F are equivalent, then $E \mid G$ should be equivalent to $F \mid G$ and $E\backslash J$ should be equivalent to $F\backslash J$, and so on for all the combinators introduced in Chapter 1. If the decision component `D` of the slot machine $\mathtt{SM_n}$ breaks down, then replacing it with an equivalent component should not affect the overall behaviour of the system (up to equivalence).

Clearly, if two processes have different initial capabilities, then they should not be deemed equivalent. Distinguishability of processes can be extended to many other features, such as their initial necessities, their (observable) traces, or their completed traces[2]. A simple technique to guarantee that an equivalence is a congruence is as follows. First, choose some simple properties as the basic distinguishable features. Second, count two processes as equivalent if, whenever they are placed in a process context, the resulting processes have the same basic properties. A process context is a process expression "with a hole in it," such as $(E \mid [\,])\backslash J$, where $[\,]$ is the "hole." This approach is sensitive to three important considerations. First is the choice of what counts as a basic distinguishable property, and whether it refers to observable behaviour as determined by the $\stackrel{a}{\Longrightarrow}$ and $\stackrel{\varepsilon}{\Longrightarrow}$ transitions, or with behaviour as presented by the single arrow transitions. Second is the choice of process operators that are permitted in the definition of a process context. Lastly, there is the question whether the resulting congruence can be characterized independently of its definition as the equivalence preserved by all process contexts.

Interesting work has been done on this topic, mostly, however, with respect to the behaviour of processes as determined by the single thin transitions $\stackrel{a}{\longrightarrow}$. Candidates for basic distinguishable features include traces and completed traces (given, respectively, by formulas of the form $\langle a_1\rangle \ldots \langle a_n\rangle \mathtt{tt}$ and $\langle a_1\rangle \ldots \langle a_n\rangle[-]\mathtt{ff}$). There are elegant results by Bloom et al, Groote and Vaandrager [7, 25, 24] that isolate congruencies for traces and completed traces. These cover very general families of process operators whose behavioural meaning is governed by the permissible format of their transition rules. The resulting conguencies are independently definable as equivalences[3]. Results for observable behaviour include those for the failures model of CSP, [31] which takes the notion of observable failure as basic.

---

[2]A trace w for E is completed if there is an F such that $E \stackrel{w}{\longrightarrow} F$ and F is unable to perform any action.

[3]They include failures equivalence (expressed in terms of formulas of the form $\langle a_1\rangle \ldots \langle a_n\rangle[K]\mathtt{ff}$), two-thirds bisimulation, two-nested simulation equivalence, and bisimulaton equivalence. Bisimulation equivalence is discussed at length in later sections.

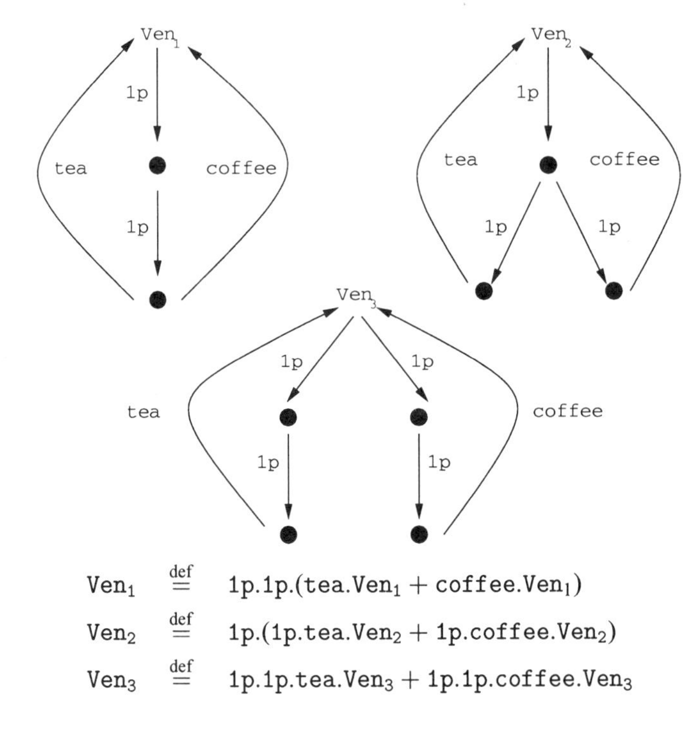

$$\begin{aligned}
\mathtt{Ven}_1 &\stackrel{\text{def}}{=} \mathtt{1p.1p.(tea.Ven}_1 + \mathtt{coffee.Ven}_1\mathtt{)} \\
\mathtt{Ven}_2 &\stackrel{\text{def}}{=} \mathtt{1p.(1p.tea.Ven}_2 + \mathtt{1p.coffee.Ven}_2\mathtt{)} \\
\mathtt{Ven}_3 &\stackrel{\text{def}}{=} \mathtt{1p.1p.tea.Ven}_3 + \mathtt{1p.1p.coffee.Ven}_3
\end{aligned}$$

FIGURE 3.2. Three vending machines

Related results are contained in the testing framework of De Nicola and Hennessy [16, 26], where processes are tested for what they may and must do.

Example 2 Consider the three similar vending machines in Figure 3.2 (where we have left out the names of the intermediate processes). These machines have the same (observable) traces. Assume a user

$$\mathtt{Use} \stackrel{\text{def}}{=} \overline{\mathtt{1p}}.\overline{\mathtt{1p}}.\overline{\mathtt{tea}}.\overline{\mathtt{ok}}.0,$$

who only wishes to drink a single tea by offering coins and, having done so, expresses visible satisfaction as the action $\overline{\mathtt{ok}}$. For each of the three vending machines, we can build the process $(\mathtt{Ven}_i \mid \mathtt{Use})\backslash K$, where K is the set $\{\mathtt{1p}, \mathtt{tea}, \mathtt{coffee}\}$. If $i = 1$ then, there is a single completed trace $\tau\ \tau\ \tau\ \overline{\mathtt{ok}}$.

$$(\mathtt{Ven}_1 \mid \mathtt{Use})\backslash K \stackrel{\tau\,\tau\,\tau\,\overline{\mathtt{ok}}}{\longrightarrow} (\mathtt{Ven}_1 \mid 0)\backslash K$$

The user must then express satisfaction after some silent activity. In the other two cases, there is another completed trace $\tau\ \tau$.

$$(\mathtt{Ven}_i \mid \mathtt{Use})\backslash K \stackrel{\tau\,\tau}{\longrightarrow} (\mathtt{coffee.Ven}_i \mid \overline{\mathtt{tea}}.\overline{\mathtt{ok}}.0)\backslash K$$

The user is then precluded from expressing satisfaction.

With respect to the failures model of CSP [31] and the testing framework of [16, 26], $\texttt{Ven}_2$ and $\texttt{Ven}_3$ are equivalent. These two processes obey the same failure formulas. Finer equivalences distinguish them on the basis that, once a coin has been inserted in $\texttt{Ven}_3$, any possible successful collection of tea is already decided. Imagine that, after a single coin has been inserted, the resulting process is copied for a number of users. In the case of $\texttt{Ven}_3$, all these users must express satisfaction, or all of them must be precluded from doing so. Let *Rep* be a process operator that replicates successor processes. There is a single transition rule for *Rep*.

$$\frac{Rep(E) \stackrel{a}{\longrightarrow} E' \mid E'}{E \stackrel{a}{\longrightarrow} E'}$$

The two processes $(Rep(\texttt{Ven}_2 \mid \texttt{Use}))\backslash K$ and $(Rep(\texttt{Ven}_3 \mid \texttt{Use}))\backslash K$ have different completed traces. The first can perform $\tau\,\tau\,\tau\,\tau\,\overline{\texttt{ok}}$ as follows, where E abbreviates $\texttt{1p.tea.Ven}_2 + \texttt{1p.coffee.Ven}_2$.

$$\begin{array}{c}
(Rep(\texttt{Ven}_2 \mid \texttt{Use}))\backslash K \\
\downarrow \tau \\
(E \mid \overline{\texttt{1p}}.\overline{\texttt{tea}}.\overline{\texttt{ok}}.0 \mid E \mid \overline{\texttt{1p}}.\overline{\texttt{tea}}.\overline{\texttt{ok}}.0)\backslash K \\
\downarrow \tau \\
(\texttt{tea.Ven}_2 \mid \overline{\texttt{tea}}.\overline{\texttt{ok}}.0 \mid E \mid \overline{\texttt{1p}}.\overline{\texttt{tea}}.\overline{\texttt{ok}}.0)\backslash K \\
\downarrow \tau \\
(\texttt{tea.Ven}_2 \mid \overline{\texttt{tea}}.\overline{\texttt{ok}}.0 \mid \texttt{coffee.Ven}_2 \mid \overline{\texttt{tea}}.\overline{\texttt{ok}}.0)\backslash K \\
\downarrow \tau \\
(\texttt{Ven}_2 \mid \overline{\texttt{ok}}.0 \mid \texttt{coffee.Ven}_2 \mid \overline{\texttt{tea}}.\overline{\texttt{ok}}.0)\backslash K \\
\downarrow \overline{\texttt{ok}} \\
(\texttt{Ven}_2 \mid 0 \mid \texttt{coffee.Ven}_2 \mid \overline{\texttt{tea}}.\overline{\texttt{ok}}.0)\backslash K
\end{array}$$

$(Rep(\texttt{Ven}_3 \mid \texttt{Use}))\backslash K$ is unable to perform this completed trace, as can be seen from its two possible initial transitions. First is the transition

$$\begin{array}{c}
(Rep(\texttt{Ven}_3 \mid \texttt{Use}))\backslash K \\
\downarrow \tau \\
(\texttt{1p.tea.Ven}_3 \mid \overline{\texttt{1p}}.\overline{\texttt{tea}}.\overline{\texttt{ok}}.0 \mid \texttt{1p.tea.Ven}_3 \mid \overline{\texttt{1p}}.\overline{\texttt{tea}}.\overline{\texttt{ok}}.0)\backslash K,
\end{array}$$

which must continue with two $\overline{\texttt{ok}}$ transitions in any completed trace. The second is

$$\begin{array}{c}
(Rep(\texttt{Ven}_3 \mid \texttt{Use}))\backslash K \\
\downarrow \tau \\
(\texttt{1p.coffee.Ven}_3 \mid \overline{\texttt{1p}}.\overline{\texttt{tea}}.\overline{\texttt{ok}}.0 \mid \texttt{1p.coffee.Ven}_3 \mid \overline{\texttt{1p}}.\overline{\texttt{tea}}.\overline{\texttt{ok}}.0)\backslash K,
\end{array}$$

which will not include $\overline{\texttt{ok}}$ transitions in any completed trace.

Exercises

1. **a.** For each $i : 1 \leq i \leq 3$, draw the transition graph of the following process $(Rep(\texttt{Ven}_i \mid \texttt{Use}_2))\backslash K$.

   b. Show the following

   i. $(Rep(\texttt{Ven}_1 \mid \texttt{Use}))\backslash K \models [\![\;]\!] \langle\!\langle - \rangle\!\rangle \texttt{tt} \wedge [\![-\overline{\texttt{ok}}]\!] \texttt{ff}$

   ii. $(Rep(\texttt{Ven}_2 \mid \texttt{Use}))\backslash K \not\models [\![\;]\!] \langle\!\langle - \rangle\!\rangle \texttt{tt} \wedge [\![-\overline{\texttt{ok}}]\!] \texttt{ff}$

   iii. $(Rep(\texttt{Ven}_2 \mid \texttt{Use}))\backslash K \models \langle\!\langle \overline{\texttt{ok}} \rangle\!\rangle [\![\overline{\texttt{ok}}]\!] \texttt{ff}$

   iv. $(Rep(\texttt{Ven}_3 \mid \texttt{Use}))\backslash K \not\models \langle\!\langle \overline{\texttt{ok}} \rangle\!\rangle [\![\overline{\texttt{ok}}]\!] \texttt{ff}$

   c. Let Γ be the set of M failure formulas $\langle a_1 \rangle \ldots \langle a_n \rangle [K]\texttt{ff}$, $n \geq 0$. Show that for all $\Phi \in \Gamma$, $\texttt{Ven}_2 \models \Phi$ iff $\texttt{Ven}_3 \models \Phi$.

2. Let $\mathrm{Tr}(E)$ be the set of traces of E (that is, the set of words $w \in \mathsf{A}^*$ such that $E \stackrel{w}{\longrightarrow} E'$ for some E'). Let $E \equiv_{\mathrm{Tr}} F$ iff $\mathrm{Tr}(E) = \mathrm{Tr}(F)$.

   a. Show that $\equiv_{\mathrm{Tr}}$ is a congruence for CCS processes.

   b. A deadlock potential for E is a trace w such that $E \stackrel{w}{\longrightarrow} E'$ and E' is unable to perform an action (so w is a completed trace). Let $\mathrm{DP}(E)$ be the set of deadlock potentials of E. Give examples of processes E and F such that $E \equiv_{\mathrm{Tr}} F$ and $\mathrm{DP}(E) \neq \mathrm{DP}(F)$.

   c. Let $E \equiv_{\mathrm{DP}} F$ iff $\mathrm{DP}(E) = \mathrm{DP}(F)$. Show that $\equiv_{\mathrm{DP}}$ is not a congruence for CCS processes.

   d. Define a process operator for which $\equiv_{\mathrm{Tr}}$ is not a congruence.

3. Design a process context that distinguishes between the two vending machines V and U, below (by having different completed traces).

$$\begin{aligned}
\mathrm{V} &\stackrel{\mathrm{def}}{=} \texttt{1p}.(\texttt{1p}.(\texttt{tea}.\mathrm{V} + \texttt{coffee}.\mathrm{V}) + \texttt{1p}.\texttt{coffee}.\mathrm{V}) \\
\mathrm{U} &\stackrel{\mathrm{def}}{=} \texttt{1p}.(\texttt{1p}.(\texttt{tea}.\mathrm{U} + \texttt{coffee}.\mathrm{U}) + \texttt{1p}.\texttt{coffee}.\mathrm{U}) + \texttt{1p}.\mathrm{U}_1 \\
\mathrm{U}_1 &\stackrel{\mathrm{def}}{=} \texttt{1p}.(\texttt{tea}.\mathrm{U} + \texttt{coffee}.\mathrm{U})
\end{aligned}$$

3.2 Interactive games

Equivalences for CCS processes begin with the idea that an observer can repeatedly interact with a process by choosing one of its available transitions. Equivalence of processes is then defined in terms of the ability of observers to match their selections so that they can proceed with additional corresponding choices. The crucial difference with the approach of the previous section is that an observer can choose a *particular* transition. Such choices cannot be directly simulated in terms of process activity[4]. These equivalences are defined in terms of bisimulation relations that capture precisely what it is for observers to match their selections. However, we first proceed with an alternative exposition using games that offer a powerful image for interaction.

The equivalence game $\mathsf{G}(E_0, F_0)$, where E_0 and F_0 are processes, is an interactive game played by two participants, players R (the refuter) and V (the verifier), who are the observers who make choices of transitions. A play of the game $\mathsf{G}(E_0, F_0)$ is a finite or infinite length sequence of pairs of proceses, $(E_0, F_0) \ldots (E_i, F_i) \ldots$. The refuter attempts to show that the initial pair (E_0, F_0) can be distinguished, whereas the verifier wishes to establish that they are equivalent. Suppose an initial part of a play is the finite sequence $(E_0, F_0) \ldots (E_j, F_j)$. The next pair (E_{j+1}, F_{j+1}) is determined by one of the following two moves.

- Player R chooses a transition $E_j \stackrel{a}{\longrightarrow} E_{j+1}$, then player V chooses a transition with the same label $F_j \stackrel{a}{\longrightarrow} F_{j+1}$
- Player R chooses a transition $F_j \stackrel{a}{\longrightarrow} F_{j+1}$, then player V chooses a transition with the same label $E_j \stackrel{a}{\longrightarrow} E_{j+1}$.

The play continues with more moves. The refuter always chooses first, then the verifier, with full knowledge of the refuter's selection, chooses a transition with the same label from the other process.

A play of a game continues until one of the players wins. As discussed in the previous section, if two processes have different initial capabilities, then they are clearly distinguishable. Consequently, any position (E_n, F_n) where one of these processes is able to carry out an initial action that the other cannot counts as a win for the refuter: that is, if there is an action $a \in \mathsf{A}$ and

$$(E_n \models \langle a \rangle \mathtt{tt} \text{ and } F_n \models [a]\mathtt{ff}) \text{ or } (E_n \models [a]\mathtt{ff} \text{ and } F_n \models \langle a \rangle \mathtt{tt}).$$

Player R can then choose a transition, and player V will be unable to match it. We call such positions "R-wins." A play is won by the refuter if it reaches an R-win position. Any other play counts as a win for player V. Consequently, the verifier wins if the play is infinite, or if the play reaches a position (E_n, F_n) and neither process has an available transition. In both these circumstances, the refuter has been unable to find a difference between the starting processes.

Example 1 The verifier wins any play of $\mathsf{G}(\mathtt{Cl}, \mathtt{Cl}_2)$. A play proceeds

$$(\mathtt{Cl}, \mathtt{Cl}_2)\,(\mathtt{Cl}, \mathtt{tick.Cl}_2)\,(\mathtt{Cl}, \mathtt{Cl}_2)\ \ldots$$

forever irrespective of the component the refuter chooses to make her move from.

[4]For instance, if E has two transitions $E \stackrel{a}{\longrightarrow} E_1$ and $E \stackrel{a}{\longrightarrow} E_2$, then the observer is able to choose either of them, but there is not a "testing" process $\overline{a}.F$ that can guarantee this choice in the context $(\overline{a}.F \mid E)\backslash\{a\}$: the two results $(F \mid E_1)\backslash\{a\}$ and $(F \mid E_2)\backslash\{a\}$ are equally likely after synchronization on a.

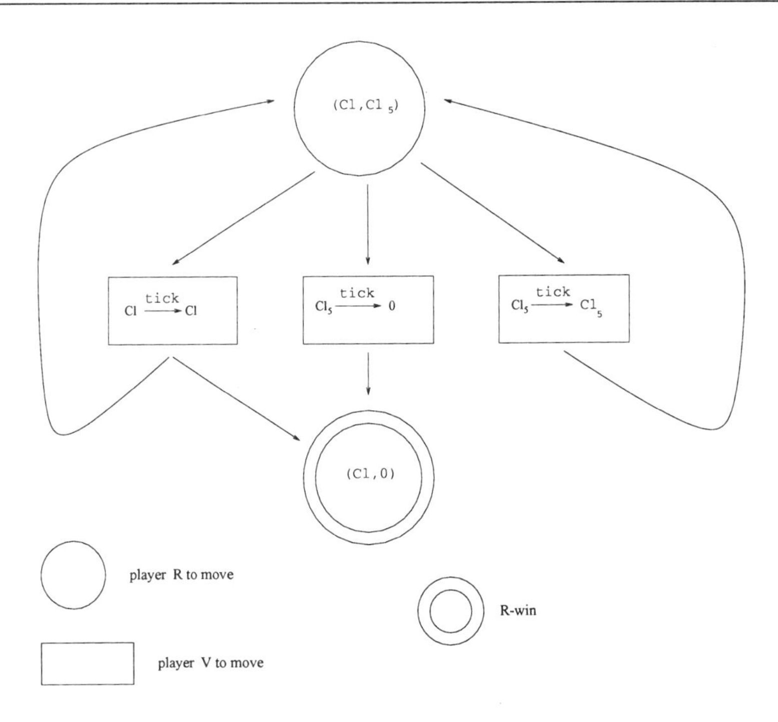

FIGURE 3.3. Game graph for $\mathsf{G}(\mathtt{Cl}, \mathtt{Cl}_5)$

Example 2 In the case of $\mathsf{G}(\mathtt{Cl}, \mathtt{Cl}_5)$, when $\mathtt{Cl}_5 \stackrel{\text{def}}{=} \mathtt{tick}.\mathtt{Cl}_5 + \mathtt{tick}.0$, there are plays that the refuter wins and plays that the verifier wins. If player R initially moves $\mathtt{Cl}_5 \stackrel{\mathtt{tick}}{\longrightarrow} 0$, then, after her opponent makes the move $\mathtt{Cl} \stackrel{\mathtt{tick}}{\longrightarrow} \mathtt{Cl}$, the resulting position $(\mathtt{Cl}, 0)$ is an R-win. If player R always chooses the transitions $\mathtt{Cl}_5 \stackrel{\mathtt{tick}}{\longrightarrow} \mathtt{Cl}_5$ or $\mathtt{Cl} \stackrel{\mathtt{tick}}{\longrightarrow} \mathtt{Cl}$, then player V can avoid defeat. Figure 3.3 depicts the game graph for $\mathsf{G}(\mathtt{Cl}, \mathtt{Cl}_5)$. Round vertices are positions from which the refuter moves, and rectangular vertices are positions from which the verifier moves. Edges of a vertex are the possible moves that a player can make from that vertex. A V-vertex is labelled with the transition player R has chosen[5]. This information constrains the choice of move that player V can make, since she must respond with a corresponding transition from the other component. Vertices encircled twice are R-wins. The game graph represents all possible plays of the game. It begins with a token on the initial vertex $(\mathtt{Cl}, \mathtt{Cl}_5)$. A play is the movement of the token around the graph. If the token is at an R-vertex, the refuter moves it, and if it is at a V-vertex the verifier moves it. If the token reaches an R-win vertex, the game stops and the refuter wins.

[5]We suppress the position in which this transition has been chosen, as can be easily seen from the graph.

Example 2 shows that different plays of a game can have different winners. Nevertheless, for each game one of the players is able to win any play irrespective of what moves her opponent makes. To make this precise, we introduce the notion of a strategy. A strategy for a player is a family of rules that tell the player how to move. For the refuter, a rule has the form "if the play so far is $(E_0, F_0) \ldots (E_i, F_i)$, then choose transition t, where t is either $E_i \xrightarrow{a_i} E_{i+1}$ or $F_i \xrightarrow{a_i} F_{i+1}$." Because the verifier responds to the refuter's choice of transition, a rule for player V has the form "if the play so far is $(E_0, F_0) \ldots (E_i, F_i)$, and player R has chosen transition t, then choose transition t'," where t' is a corresponding transition of the other process. However, it turns out that we only need to consider history-free strategies whose rules do not depend upon previous positions in the play. For player R, a rule is therefore of the form

at position (E, F) choose transition t,

where t is either $E \xrightarrow{a} E'$ or $F \xrightarrow{a} F'$. A rule for player V is

at position (E, F) when playerRhaschosen t choose t',

where t is either $E \xrightarrow{a} E'$ or $F \xrightarrow{a} F'$ and t' is a corresponding transition of the other process. A player uses a strategy in a play if all her moves obey the rules in it. A strategy is a winning one if the player wins every play in which she uses it. If a player has a winning strategy for a game, we say that the player "wins the game."

Example 3 Player R's winning strategy for the game G($\mathtt{Cl}$, $\mathtt{Cl}_5$) consists of the single rule "at ($\mathtt{Cl}$, $\mathtt{Cl}_5$) choose $\mathtt{Cl}_5 \xrightarrow{\mathtt{tick}} \mathtt{0}$." This has the effect of reducing the game graph of Figure 3.3 to the smaller subgraph in Figure 3.4, as redundant player R choices are removed.

Example 4 The game graph for G($\mathtt{Ven}_2$, $\mathtt{Ven}_3$), two of the vending machines of Figure 3.2, is pictured in Figure 3.5. A winning strategy for the refuter consists of the following two rules (where E, G and H are from Figure 3.5).

at ($\mathtt{Ven}_2$, $\mathtt{Ven}_3$) choose $\mathtt{Ven}_3 \xrightarrow{\mathtt{1p}} G$

at (E, G) choose $E \xrightarrow{\mathtt{1p}} H$

The reader is invited to find another winning strategy.

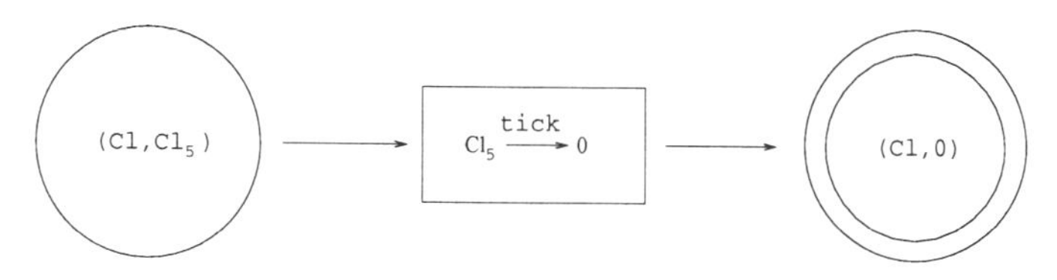

FIGURE 3.4. Reduced game graph for G($\mathtt{Cl}$, $\mathtt{Cl}_5$)

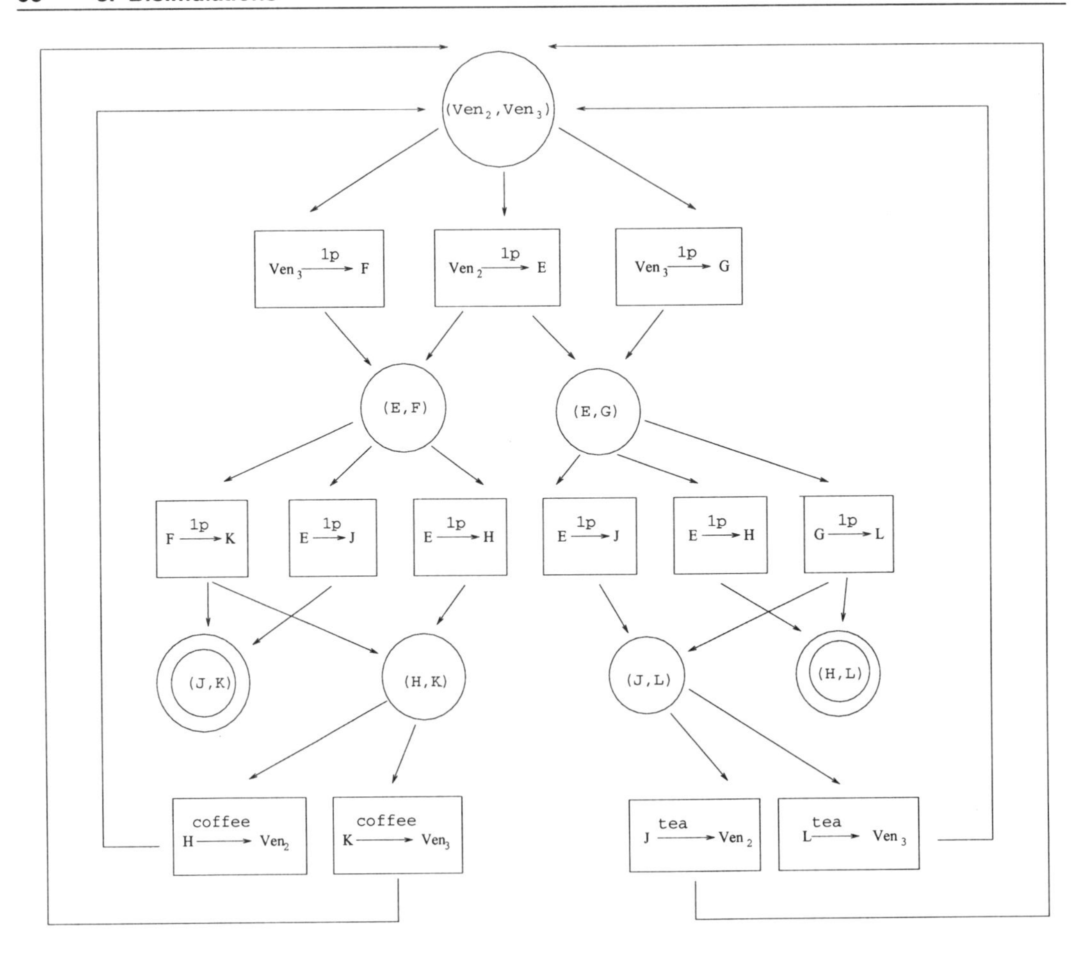

$$E \equiv \mathtt{1p.tea.Ven_2} + \mathtt{1p.coffee.Ven_2}$$

$$\begin{array}{lllll} F & \equiv & \mathtt{1p.coffee.Ven_3} & G & \equiv & \mathtt{1p.tea.Ven_3} \\ H & \equiv & \mathtt{coffee.Ven_2} & J & \equiv & \mathtt{tea.Ven_2} \\ K & \equiv & \mathtt{coffee.Ven_3} & L & \equiv & \mathtt{tea.Ven_3} \end{array}$$

FIGURE 3.5. Game graph for ($\mathtt{Ven_2}$, $\mathtt{Ven_3}$)

For each game, one of the players has a winning strategy. This we shall prove below. The strategy relies on defining sets of pairs of processes iteratively using ordinals as indices. Ordinals are ordered in increasing size, as follows.

$$0, 1, \ldots, \omega, \omega + 1, \ldots, \omega + \omega, \omega + \omega + 1, \ldots$$

The initial limit ordinal (that is, one without an immediate predecessor) is ω and $\omega + 1$ is its successor. The next limit ordinal is $\omega + \omega$.

Theorem 1 *For any game* $\mathsf{G}(E, F)$*, either player R or player V has a history-free winning strategy.*

Proof. Consider the game $\mathsf{G}(E, F)$. The set of possible player R positions P is the set

$$P = \{(E', F') : \exists w \in \mathsf{A}^*.\, E \xrightarrow{w} E' \text{ and } F \xrightarrow{w} F'\},$$

where $\xrightarrow{w}$ is the extended transition relation defined in Section 1.3. P contains all possible positions of the game $\mathsf{G}(E, F)$ in which player R moves next. Let $W \subseteq P$ be the subset of positions that are R-wins. We now define the subset of positions from which player R can force a win by eventually entering W. This set Force is defined iteratively, starting with 1 and using ordinals, where λ is a limit ordinal.

$$\begin{aligned}
\text{Force}^1 &= W \\
\text{Force}^{\alpha+1} &= \text{Force}^\alpha \cup \{(G, H) \in P : \\
&\quad \exists G'.\, G \xrightarrow{a} G' \text{ and } \forall H'.\text{ if } H \xrightarrow{a} H' \text{ then } (G', H') \in \text{Force}^\alpha \\
&\quad \text{or } \exists H'.\, H \xrightarrow{a} H' \text{ and } \forall G'.\text{ if } G \xrightarrow{a} G' \text{ then } (G', H') \in \text{Force}^\alpha\} \\
\text{Force}^\lambda &= \bigcup\{\text{Force}^\alpha : \alpha < \lambda\}
\end{aligned}$$

Lastly we define[6] Force as the following subset of positions.

$$\text{Force} = \bigcup\{\text{Force}^\alpha : \alpha > 0\}$$

If $(G, H) \in$ Force, then the rank of (G, H) is the least ordinal α such that $(G, H) \in \text{Force}^\alpha$. For each $(G, H) \in$ Force, player R has a history-free winning strategy for the game $\mathsf{G}(G, H)$. The strategy consists of rules of the form, "if (E', F') has rank $\alpha > 1$, then choose transition t such that, whatever choice of transition player V makes, the resulting pair of processes has lower rank." The definition of Force guarantees that there is a choice of transition with this property.

If $(G, H) \notin$ Force, then player V has a history-free winning strategy, which is to avoid the set Force. It consists of rules of the form, "if $(E', F') \notin$ Force and player R chooses t, then choose t' such that the resulting pair of processes does not belong to Force." The initial pair of processes (E, F) either belongs to Force or belongs to P − Force, meaning one of the players has a history-free winning strategy for the game $\mathsf{G}(E, F)$. □

If player V wins the game $\mathsf{G}(E, F)$, then we say that process E is "game equivalent" to process F, in which case player R cannot detect a difference in behaviour between the processes E and F. Game equivalence is indeed an equivalence relation.

---

[6] The processes considered in this work have countable transition graphs, so the set P of positions is countable; therefore, we need only consider ordinals whose cardinality is at most that of $\mathbb{N}$.

Proposition 1 *Game equivalence between processes is an equivalence relation.*

Proof. We show that game equivalence is an equivalence relation: that is, we show it is reflexive (E is equivalent to E), symmetric (if E is equivalent to F, then F is equivalent to E) and transitive (if E is equivalent to F and F is equivalent to G, then E is equivalent to G).

Player V's winning strategy for $\mathsf{G}(E, E)$ is the "copy-cat strategy" consisting of the rules "at (F, F), when player R has chosen t, choose t." For symmetry, suppose that π is a history-free winning strategy for player V for the game $\mathsf{G}(E, F)$. Let π' be the symmetric strategy that changes each rule "at (G, H) ..." in π to "at (H, G) ...". Clearly, π' is a history-free winning strategy for player V for the game $\mathsf{G}(F, E)$. Next, assume σ is a winning strategy for player V for $\mathsf{G}(E, F)$, and π is a winning strategy for player V for $\mathsf{G}(F, G)$. The composition of these strategies, $\pi \circ \sigma$, is a winning strategy for player V for $\mathsf{G}(E, G)$. Composition is defined by the following two closure conditions.

1. If "at (E', F'), when player R has chosen $E' \stackrel{a}{\longrightarrow} E''$, choose t'" is in σ, and "at (F', G'), when player R has chosen t', choose t" is in π, then "at (E', G'), when player R has chosen $E' \stackrel{a}{\longrightarrow} E''$, choose t" is in $\pi \circ \sigma$
2. If "at (F', G'), when player R has chosen $G' \stackrel{a}{\longrightarrow} G''$, choose t'" is in π, and "at (E', F'), when player R has chosen t', choose t" is in σ, then "at (E', G'), when player R has chosen $G' \stackrel{a}{\longrightarrow} G''$, choose t" is in $\pi \circ \sigma$

We leave as an exercise that $\pi \circ \sigma$ is a winning strategy for player V for the game $\mathsf{G}(E, G)$. □

If E and F are finite state processes, then the proof of Theorem 1 provides a straightforward algorithm for deciding whether E is game equivalent to F. Assume that the number of processes in the transition graphs for both E and F are at most n. One first computes the set P of possible player R positions, $\{(E', F') : E \stackrel{w}{\longrightarrow} E' \text{ and } F \stackrel{w}{\longrightarrow} F'\}$. The size of this set is therefore bounded by n^2. One then picks out Force^1, the subset $W \subseteq P$ of R-wins. Next, one defines iteratively the sets Force^{i+1} for $i \geq 1$, by adding pairs from $P - \text{Force}^i$ obeying the requirement. The algorithm stops as soon as the set Force^i is the same set as Force^{i-1}. This set is then the set Force. It is clear that there can be at most n^2 iterations before this happens. If $(E, F) \notin$ Force, then E is game equivalent to F, otherwise, they are not game equivalent.

Exercises

1. Draw the game graphs for $\mathsf{G}(\mathtt{Ven}_1, \mathtt{Ven}_2)$ and $\mathsf{G}(\mathtt{Ven}_1, \mathtt{Ven}_3)$ where the vending machines are defined in Figure 3.2.

2. Show that the pair of vending machines V and U, below, are not game equivalent.

$$\begin{aligned} \mathtt{V} &\stackrel{\text{def}}{=} \mathtt{1p.(1p.(tea.V + coffee.V) + 1p.coffee.V)} \\ \mathtt{U} &\stackrel{\text{def}}{=} \mathtt{1p.(1p.(tea.U + coffee.U) + 1p.coffee.U) + 1p.U_1} \\ \mathtt{U_1} &\stackrel{\text{def}}{=} \mathtt{1p.(tea.U + coffee.U)} \end{aligned}$$

3. Show that Player V has a winning strategy for the game

$$\mathsf{G}((\mathtt{C} \mid \mathtt{U})\backslash\{\mathtt{in}, \mathtt{ok}\}, \mathtt{Ucop}'),$$

where these processes are

$$\begin{aligned} \mathtt{C} &\stackrel{\text{def}}{=} \mathtt{in}(x).\overline{\mathtt{out}}(x).\overline{\mathtt{ok}}.\mathtt{C} \\ \mathtt{U} &\stackrel{\text{def}}{=} \mathtt{write}(x).\overline{\mathtt{in}}(x).\mathtt{ok}.\mathtt{U} \\ \mathtt{Ucop}' &\stackrel{\text{def}}{=} \mathtt{write}(x).\tau.\overline{\mathtt{out}}(x).\tau.\mathtt{Ucop}'. \end{aligned}$$

4. Consider adding the following winning condition for player V: If a position is repeated (occurs earlier in a play), then player V wins the play.

   a. Show that this extra winning condition does not affect which player wins a game $\mathsf{G}(E, F)$.

   b. Find out what it means for a problem to be **P**-complete. For example, see Papadimitriou [48].

   c. Show that game equivalence between finite state processes E and F is **P**-complete.

5. A process F is immediately image-finite if, for each $a \in \mathsf{A}$, the set $\{G : F \stackrel{a}{\longrightarrow} G\}$ is finite. F is image-finite if each F' is immediately image-finite whenever $F \stackrel{w}{\longrightarrow} F'$ for $w \in \mathsf{A}^*$.

   a. Show that, if the starting processes E and F of Theorem 1 are image-finite, then for the proof of the result it suffices to define Force as follows.

$$\text{Force} = \bigcup \{\text{Force}^i : i \in \mathbb{N}\}$$

   b. Give an example of a pair of processes (E, F) that are not game equivalent, but that fail to have rank i for any $i \in \mathbb{N}$.

6. A pair of processes (E_0, F_0) is an S-game if, in any play, player R must always choose a transition from the left process, meaning only the first move is allowed. An R-win is any position (E_n, F_n) such that $E_n \stackrel{a}{\longrightarrow} E_{n+1}$, but F_n has no available a transition. Player V wins if the play does not reach an R-win.

   a. Show that, for each S-game, one of the players has a history-free winning strategy.

b. List all the pairs of vending machines from $\mathtt{Ven}_1$, $\mathtt{Ven}_2$, and $\mathtt{Ven}_3$ of the previous section for which player V has a winning strategy for the S-game.

c. E is S-equivalent to F if player V has a winning strategy for the two S-games (E, F) and (F, E). Prove that S-equivalence is an equivalence relation. Give an example of two processes that are S-equivalent, but not game equivalent.

3.3 Bisimulation relations

When E and F are game equivalent, player V can match player R's choice of transition: if $E \xrightarrow{a} E'$, then there is a transition $F \xrightarrow{a} F'$ such that E' and F' are also game equivalent, and if $F \xrightarrow{a} F'$, then again there is a transition $E \xrightarrow{a} E'$ such that E' and F' are game equivalent. The ability to match transitions is defining for a bisimulation relation. Bisimulations were introduced[7] by Park [47] as a refinement of the iteratively defined equivalence of Hennessy and Milner [29, 42].

Definition 1 A binary relation B between processes is a bisimulation provided that, whenever $(E, F) \in B$ and $a \in \mathsf{A}$,

- if $E \xrightarrow{a} E'$ then $F \xrightarrow{a} F'$ for some F' such that $(E', F') \in B$, and
- if $F \xrightarrow{a} F'$ then $E \xrightarrow{a} E'$ for some E' such that $(E', F') \in B$

A binary relation between processes is a bisimulation provided that it obeys the two hereditary conditions in the definition. Simple examples of bisimulations are the identity relation and the empty relation.

Example 1 Assume $\mathtt{Cl}$, $\mathtt{Cl}_2$ and $\mathtt{Cl}_5$ are the clocks of the previous section. The relation $B = \{(\mathtt{Cl}, \mathtt{Cl}_2), (\mathtt{Cl}, \mathtt{tick}.\mathtt{Cl}_2)\}$ is a bisimulation. For example, if $\mathtt{Cl} \xrightarrow{\mathtt{tick}} \mathtt{Cl}$ then $\mathtt{Cl}_2 \xrightarrow{\mathtt{tick}} \mathtt{tick}.\mathtt{Cl}_2$ and the resulting pair of processes belongs to B. The relation $\{(\mathtt{Cl}, \mathtt{Cl}_5)\}$ is *not* a bisimulation because of the transition $\mathtt{Cl}_5 \xrightarrow{\mathtt{tick}} 0$. Adding $(\mathtt{Cl}, 0)$ does not rectify this because the transition $\mathtt{Cl} \xrightarrow{\mathtt{tick}} \mathtt{Cl}$ cannot then be matched by a transition of the process 0.

Two processes E and F are bisimulation equivalent (or bisimilar) if there is a bisimulation relation B such that $(E, F) \in B$. We write $E \sim F$ if E and F are bisimilar.

---

[7] They also occur in a slightly different form in the theory of modal logic as zig-zag relations; see Benthem [6].

Example 2 The following processes are not bisimilar, $a.(b.0 + c.0)$ and $a.b.0 + a.c.0$. There cannot be a bisimulation relating the pair because it would have to include either $(b.0 + c.0, b.0)$ or $(b.0 + c.0, c.0)$.

Proposition 1 *If* $\{B_i : i \in I\}$ *is a family of bisimulations, then their union* $\bigcup\{B_i : i \in I\}$ *is a bisimulation.*

Proof. Let B be the relation $\bigcup\{B_i : i \in I\}$, and suppose $(E, F) \in B$. Therefore, $(E, F) \in B_j$ for some $j \in I$. If $E \xrightarrow{a} E'$, then $F \xrightarrow{a} F'$ for some F' and $(E', F') \in B_j$, and similarly if $F \xrightarrow{a} F'$, then $E \xrightarrow{a} E'$ for some E' and $(E', F') \in B_j$. Therefore, in both cases $(E', F') \in B$. □

A corollary of Proposition 1 is that the binary relation $\sim$ is itself a bisimulation because it is defined as $\bigcup\{B : B \text{ is a bisimulation}\}$. Consequently, $\sim$ is the largest bisimulation (with respect to subset inclusion).

Bisimulation equivalence and game equivalence coincide.

Proposition 2 *E is game equivalent to F iff* $E \sim F$.

Proof. Assume that E is game equivalent to F. We show that $E \sim F$ by establishing that the relation $B = \{(E', F') : E' \text{ and } F' \text{ are game equivalent}\}$ is a bisimulation. Suppose $E' \xrightarrow{a} E''$, and because this is a possible move by player R, we know that player V can respond with $F' \xrightarrow{a} F''$ in such a way that $(E'', F'') \in B$, and similarly for a player R move $F' \xrightarrow{a} F''$. For the other direction, suppose $E \sim F$, so there is a bisimulation relation B such that $(E, F) \in B$. We construct a winning strategy for player V for the game $\mathsf{G}(E, F)$. The idea is that in any play, whatever move player R makes, player V responds with a move ensuring that the resulting pair of processes remains in the relation B. So, the winning strategy for the verifier consists of rules of the form "if $(E', F') \in B$ when R has chosen $E' \xrightarrow{a} E''$, then choose $F' \xrightarrow{a} F''$ such that $(E'', F'') \in B$" and similarly for the case when the refuter chooses transitions from the other component. □

The parallel operator $|$ and the sum operator $+$ are both commutative and associative with respect to bisimulation equivalence. This means that the following hold for arbitrary processes E, F and G.

$$\begin{array}{lllllll} E \mid F & \sim & F \mid E & \quad & (E \mid F) \mid G & \sim & E \mid (F \mid G) \\ E + F & \sim & F + E & \quad & (E + F) + G & \sim & E + (F + G) \end{array}$$

This is further justification for dropping brackets in the case of a process description having multiple parallel components, or multiple sum components, such as the description of `Crossing` in Section 1.2.

To show that two processes are bisimilar, it is sufficient to exhibit a bisimulation that contains them. This offers a very straightforward proof technique for bisimilarity.

Example 3 The two processes $\mathtt{Cnt}$ and $\mathtt{Ct}'_0$ are bisimilar where

$$\begin{aligned}
\mathtt{Cnt} &\stackrel{\text{def}}{=} \mathtt{up}.(\mathtt{Cnt} \mid \mathtt{down}.0) \\
\mathtt{Ct}'_0 &\stackrel{\text{def}}{=} \mathtt{up}.\mathtt{Ct}'_1 \\
\mathtt{Ct}'_{i+1} &\stackrel{\text{def}}{=} \mathtt{up}.\mathtt{Ct}'_{i+2} + \mathtt{down}.\mathtt{Ct}'_i \;\; i \geq 0.
\end{aligned}$$

A bisimulation that contains the pair $(\mathtt{Cnt}, \mathtt{Ct}'_0)$ has infinite size because the processes are infinite state. Let P_i be the following families of processes for $i \geq 0$ (when brackets are dropped between parallel components)

$$\begin{aligned}
P_0 &= \{\mathtt{Cnt} \mid 0^j \; : \; j \geq 0\} \\
P_{i+1} &= \{E \mid 0^j \mid \mathtt{down}.0 \mid 0^k \; : \; E \in P_i \text{ and } j \geq 0 \text{ and } k \geq 0\},
\end{aligned}$$

where $F \mid 0^0 = F$ and $F \mid 0^{i+1} = F \mid 0^i \mid 0$. The following relation

$$B = \{(E, \mathtt{Ct}'_i) \; : \; i \geq 0 \text{ and } E \in P_i\}$$

is a bisimulation that contains the pair $(\mathtt{Cnt}, \mathtt{Ct}'_0)$. The proof that it is a bisimulation proceeds by case analysis. If $i = 0$, then $(\mathtt{Cnt} \mid 0^j, \mathtt{Ct}'_0) \in B$ for any $j \geq 0$. Because $\mathtt{Cnt} \xrightarrow{\mathtt{up}} \mathtt{Cnt} \mid \mathtt{down}.0$, it follows that $\mathtt{Cnt} \mid 0^j \xrightarrow{\mathtt{up}} \mathtt{Cnt} \mid \mathtt{down}.0 \mid 0^j$. This transition is matched by $\mathtt{Ct}'_0 \xrightarrow{\mathtt{up}} \mathtt{Ct}'_1$ because $\mathtt{Cnt} \mid \mathtt{down}.0 \mid 0^j \in P_1$. The other case when $i > 0$ is left as an exercise for the reader.

Example 4 Assume that $\mathtt{C}$ and $\mathtt{U}$ are as follows.

$$\begin{aligned}
\mathtt{C} &\stackrel{\text{def}}{=} \mathtt{in}(x).\overline{\mathtt{out}}(x).\overline{\mathtt{ok}}.\mathtt{C} \\
\mathtt{U} &\stackrel{\text{def}}{=} \mathtt{write}(x).\overline{\mathtt{in}}(x).\mathtt{ok}.\mathtt{U}
\end{aligned}$$

The proof that $(\mathtt{C} \mid \mathtt{U})\backslash\{\mathtt{in}, \mathtt{ok}\} \sim (\mathtt{C} \mid \mathtt{C} \mid \mathtt{U})\backslash\{\mathtt{in}, \mathtt{ok}\}$ is given by the following bisimulation relation B.

$$\begin{aligned}
&\{((\mathtt{C} \mid \mathtt{U})\backslash\{\mathtt{in}, \mathtt{ok}\}, (\mathtt{C} \mid \mathtt{C} \mid \mathtt{U})\backslash\{\mathtt{in}, \mathtt{ok}\})\} \cup \\
&\{((\mathtt{C} \mid \overline{\mathtt{in}}(v).\mathtt{ok}.\mathtt{U})\backslash\{\mathtt{in}, \mathtt{ok}\}, (\mathtt{C} \mid \mathtt{C} \mid \overline{\mathtt{in}}(v).\mathtt{ok}.\mathtt{U})\backslash\{\mathtt{in}, \mathtt{ok}\}) \; : \; v \in D\} \cup \\
&\{((\overline{\mathtt{out}}(v).\overline{\mathtt{ok}}.\mathtt{C} \mid \mathtt{ok}.\mathtt{U})\backslash\{\mathtt{in}, \mathtt{ok}\}, (\overline{\mathtt{out}}(v).\overline{\mathtt{ok}}.\mathtt{C} \mid \mathtt{C} \mid \mathtt{ok}.\mathtt{U})\backslash\{\mathtt{in}, \mathtt{ok}\}) \; : \; v \in D\} \cup \\
&\{((\overline{\mathtt{out}}(v).\overline{\mathtt{ok}}.\mathtt{C} \mid \mathtt{ok}.\mathtt{U})\backslash\{\mathtt{in}, \mathtt{ok}\}, (\mathtt{C} \mid \overline{\mathtt{out}}(v).\overline{\mathtt{ok}}.\mathtt{C} \mid \mathtt{ok}.\mathtt{U})\backslash\{\mathtt{in}, \mathtt{ok}\}) \; : \; v \in D\} \cup \\
&\{((\overline{\mathtt{ok}}.\mathtt{C} \mid \mathtt{ok}.\mathtt{U})\backslash\{\mathtt{in}, \mathtt{ok}\}, (\overline{\mathtt{ok}}.\mathtt{C} \mid \mathtt{C} \mid \mathtt{ok}.\mathtt{U})\backslash\{\mathtt{in}, \mathtt{ok}\}) \; : \; v \in D\} \cup \\
&\{((\overline{\mathtt{ok}}.\mathtt{C} \mid \mathtt{ok}.\mathtt{U})\backslash\{\mathtt{in}, \mathtt{ok}\}, (\mathtt{C} \mid \overline{\mathtt{ok}}.\mathtt{C} \mid \mathtt{ok}.\mathtt{U})\backslash\{\mathtt{in}, \mathtt{ok}\}) \; : \; v \in D\}
\end{aligned}$$

We leave the reader to check that B is indeed a bisimulation.

Bisimulation equivalence is also a congruence with respect to all the process combinators introduced in previous sections (including the operator *Rep*).

Proposition 3 *If $E \sim F$, then for any process G, for any set of actions K, for any action a and for any renaming function f,*

1. $a.E \sim a.F$	2. $E+G \sim F+G$	3. $E \mid G \sim F \mid G$
4. $E[f] \sim F[f]$	5. $E\backslash K \sim F\backslash K$	6. $E\backslash\backslash K \sim F\backslash\backslash K$
7. $E\|_K G \sim F\|_K G$	8. $Rep(E) \sim Rep(F)$.	

Proof. We show case 3 and leave the other cases for the reader to prove. The relation $B = \{(E \mid G, F \mid G) : E \sim F\}$ is a bisimulation. Assume that $((E \mid G), (F \mid G)) \in B$ and $E \mid G \stackrel{a}{\longrightarrow} E' \mid G'$. There are three possibilities. First, $E \stackrel{a}{\longrightarrow} E'$ and $G = G'$. Because $E \sim F$, we know that $F \stackrel{a}{\longrightarrow} F'$ and $E' \sim F'$ for some F'. Therefore $F \mid G \stackrel{a}{\longrightarrow} F' \mid G$, and so by definition $((E' \mid G), (F' \mid G)) \in B$. Next, suppose $G \stackrel{a}{\longrightarrow} G'$ and $E' = E$. So $F \mid G \stackrel{a}{\longrightarrow} F \mid G'$, and by definition $((E \mid G'), (F \mid G')) \in B$. The last case is that $E \mid G \stackrel{\tau}{\longrightarrow} E' \mid G'$ and $E \stackrel{a}{\longrightarrow} E'$ and $G \stackrel{\overline{a}}{\longrightarrow} G'$. However, $F \stackrel{a}{\longrightarrow} F'$ for some F' such that $E' \sim F'$, so $F \mid G \stackrel{\tau}{\longrightarrow} F' \mid G'$, and therefore $((E' \mid G'), (F' \mid G')) \in B$. The argument is symmetric for a transition $F \mid G \stackrel{a}{\longrightarrow} F' \mid G'$. □

Bisimulation equivalence is a very fine equivalence between processes, reflecting the fact that, in the presence of concurrency, a more intensional description of process behaviour is needed than, for instance its set of traces. For full CCS, the question whether two processes are bisimilar is undecidable. As was mentioned in Section 1.6 Turing machines can be "coded" in CCS. Let $\mathtt{TM}_n$ be this coding of the nth Turing machine when all observable actions are hidden (using \\, which can be defined in CCS). The undecidable Turing machine halting problem is equivalent to whether $\mathtt{TM}_n \sim \mathtt{Div}$, where $\mathtt{Div} \stackrel{\text{def}}{=} \tau.\mathtt{Div}$. However, an interesting question is for what subclasses of processes it is decidable? Clearly, this is the case for finite state processes, since there are only finitely many candidates for being a bisimulation. Surprisingly, it is also decidable for families of infinite state processes including "context-free processes"[8] for which other equivalences are undecidable; see the survey by Hirshfeld and Moller [30]. One can also show decidability of bisimilarity for various classes of value passing processes whose data may be drawn from an infinite value space; see Hennessy and Lin [28].

Exercises

1. Complete the proof that the relation B in example 3 is a bisimulation.
2. Show that the relation B of example 4 is a bisimulation.
3. Prove directly that $\sim$ is an equivalence relation.
4. Assume that processes C and U are as in example 4. Show the following

   $$(\mathtt{C} \mid \mathtt{U})\backslash\{\mathtt{in}, \mathtt{ok}\} \sim (\mathtt{C}^n \mid \mathtt{U})\backslash\{\mathtt{in}, \mathtt{ok}\}$$

   for all $n \geq 1$, where $\mathtt{C}^1 = \mathtt{C}$ and $\mathtt{C}^{i+1} = \mathtt{C}^i \mid \mathtt{C}$.
5. Suppose that B and S are bisimulations. For each of the following, either prove that it is true, or provide a counterexample.

[8] These are the family of processes given by context-free grammars.

a. B^{-1} is a bisimulation

b. $B \cap S$ is a bisimulation

c. $B \cup S$ is a bisimulation

d. $-B$ is a bisimulation

e. $B \circ S$ is a bisimulation

where

$$B^{-1} = \{(F, E) : (E, F) \in B\}$$
$$-B = \{(E, F) : (E, F) \notin B\}$$
$$B \circ S = \{(E, G) : \text{there is an } F.\ (E, F) \in B \text{ and } (F, G) \in S\}$$

6. Prove the remaining cases of Proposition 3.

7. Define a process operator for which bisimulation equivalence is not a congruence.

8. A relation B between processes is a simulation (half of a bisimulation) provided that, whenever $(E, F) \in B$ and $a \in \mathsf{A}$, if $E \xrightarrow{a} E'$, then $F \xrightarrow{a} F'$ and $(E', F') \in B$ for some F'. E and F are simulation equivalent provided that there are simulations B and S such that $(E, F) \in B$ and $(F, E) \in S$.

a. List all the pairs (E, F) of vending machines from $\texttt{Ven}_1$, $\texttt{Ven}_2$, and $\texttt{Ven}_3$ of Figure 3.2 for which there is a simulation B containing (E, F).

b. Give an example of two processes which are simulation equivalent but, not bisimilar.

c. Show that E and F are simulation equivalent if, and only if, they are S-equivalent (defined in the exercises of the previous section).

9. For each ordinal α, the notion of α-equivalence, $\sim_\alpha$, is defined as follows. First, the base case, $E \sim_0 F$ for all E and F. Next, for a successor ordinal, $E \sim_{\alpha+1} F$ iff for any $a \in A$,

$$\text{if } E \xrightarrow{a} E' \text{ then for some } F'.\ F \xrightarrow{a} F' \text{ and } E' \sim_\alpha F'$$
$$\text{if } F \xrightarrow{a} F' \text{ then for some } E'.\ E \xrightarrow{a} E' \text{ and } E' \sim_\alpha F'.$$

Lastly, for a limit ordinal λ, $E \sim_\lambda F$ if, and only if, $E \sim_\alpha F$ for all $\alpha < \lambda$.

a. Give an example of a pair of processes E, F such that $E \sim_3 F$ but $E \not\sim_4 F$.

b. Consider the game $\mathsf{G}(E, F)$ as defined in the previous section. Show that, for any possible player R position, (G, H) of this game, $G \not\sim_\alpha H$ iff $(G, H) \in \text{Force}^\alpha$.

c. Prove that $E \sim F$ iff $E \sim_\alpha F$ for all α.

d. E is image-finite if for every word w the set $\{E' : E \xrightarrow{w} E'\}$ has finite size. Show that, if E and F are image-finite, then $E \sim F$ iff $E \sim_n F$ for all $n \geq 0$.

e. Give an example of a pair of processes for which $E \not\sim F$, but $E \sim_n F$ for all $n \geq 0$.

10. Suppose E and F are finite state processes. Using the notion of α-equivalence of the previous exercise, design efficient algorithms

a. for determining whether $E \sim F$. (Hint: define a least function $f : \mathbb{N} \times \mathbb{N} \rightarrow \mathbb{N}$ such that $E \sim F$ iff $E \sim_{f(|E|,|F|)} F$.)

b. which also present a bisimulation relation containing (E, F) when $E \sim F$.

How do your algorithms compare with those presented in Kannellakis and Smolka [33]?

3.4 Modal properties and equivalences

An alternative approach to defining equivalence between processes uses properties. Two processes are equivalent if they share the same properties. To understand this further, we need an accounting of properties. In Chapter 2, we introduced a variety of modal logics for describing properties of processes. Therefore, we can use modal formulas as properties. If Γ is a set of modal formulas, then the equivalence $\equiv_\Gamma$ between processes, meaning "sharing the same Γ properties," is defined as follows[9].

$$E \equiv_\Gamma F \text{ iff } \{\Phi \in \Gamma : E \models \Phi\} = \{\Psi \in \Gamma : F \models \Psi\}$$

One extreme case is when Γ is the empty set, and then $\equiv_\Gamma$ relates all pairs of processes. Families of formulas provide the basis for defining various equivalences. For instance, if Γ consists of all formulas of the form $\langle a_1 \rangle \ldots \langle a_n \rangle \mathtt{tt}$ for $n \geq 0$, the relation $\equiv_\Gamma$ is trace equivalence. Similarly, if Γ consists of formulas $\langle a_1 \rangle \ldots \langle a_n \rangle [-]\mathtt{ff}$, for all $n \geq 0$ the induced equivalence is completed trace equivalence. To capture observable equivalences, one uses subsets of M^o formulas.

As remarked above, $\equiv_\emptyset$ relates all processes. The other extreme is when Γ consists of *all* modal formulas defined in Chapter 2. More generally, let Γ be the set of formulas built from the constants $\mathtt{tt}$ and $\mathtt{ff}$, the boolean connectives $\wedge$ and $\vee$ and modal operators $[K]$, $\langle K \rangle$, $[\![-]\!]$, $\langle\!\langle - \rangle\!\rangle$, $[\![\downarrow]\!]$ and $\langle\!\langle \uparrow \rangle\!\rangle$. Γ encompasses all the modal logics defined in Chapter 2. It turns out that bisimilar processes have the same modal properties.

Proposition 1 *If $E \sim F$, then $E \equiv_\Gamma F$.*

Proof. By induction on modal formulas Φ, we show that, for any G and H, if $G \sim H$, then $G \models \Phi$ iff $H \models \Phi$. The base case when Φ is $\mathtt{tt}$ or $\mathtt{ff}$ is

[9] Similarly a "refinement" preorder $\sqsubseteq_\Gamma$ is definable as follows: $E \sqsubseteq_\Gamma F$ iff for all $\Phi \in \Gamma$. if $E \models \Phi$ then $F \models \Phi$.

clear. For the inductive step, the proof proceeds by case analysis. First, let Φ be $\Psi_1 \wedge \Psi_2$ and assume that the result holds for Ψ_1 and Ψ_2. By the definition of the satisfaction relation $G \models \Phi$ iff $G \models \Psi_1$ and $G \models \Psi_2$ iff by the induction hypothesis $H \models \Psi_1$ and $H \models \Psi_2$, and therefore iff $H \models \Phi$. A similar argument applies to $\Psi_1 \vee \Psi_2$. Next, assume that Φ is $[K]\Psi$ and $G \models \Phi$. Therefore, for any G' such that $G \xrightarrow{a} G'$ and $a \in K$, it follows that $G' \models \Psi$. To show that $H \models \Phi$, let $H \xrightarrow{a} H'$ (with $a \in K$). However, we know that for some G' there is the transition $G \xrightarrow{a} G'$ and $G' \sim H'$, so by the induction hypothesis $H' \models \Psi$, and therefore $H \models \Phi$. The other modal cases are left as an exercise for the reader. □

This result tells us that two bisimilar processes have the same capabilities, the same necessities and the same divergence potentials. Although the converse of Proposition 1 does not hold in general (see Example 1, below), it does hold in the case of a restricted set of processes. A process E is immediately image-finite if, for each $a \in \mathsf{A}$, the set $\{F : E \xrightarrow{a} F\}$ is finite. For each $a \in \mathsf{A}$, E has only finitely many a-transitions. E is image-finite if every member of $\{F : \exists w \in \mathsf{A}^*.\ E \xrightarrow{w} F\}$ is immediately image-finite. That is, a process is image-finite if all processes in its transition graph are immediately image-finite. With this restriction to image-finite processes, the converse of Proposition 1 holds.

Proposition 2 *If E and F are image-finite and $E \equiv_\Gamma F$, then $E \sim F$.*

Proof. It suffices to prove the result for the case when Γ is the set of modal formulas M of Section 2.1. We show that the following relation is a bisimulation.

$$\{(E, F) : E \equiv_{\mathrm{M}} F \text{ and } E, F \text{ are image} - \text{finite}\}$$

But suppose not. Therefore, without loss of generality $G \equiv_{\mathrm{M}} H$ for some G and H, and $G \xrightarrow{a} G'$ for some a and G', but $G' \not\equiv_{\mathrm{M}} H'$ for all H' such that $H \xrightarrow{a} H'$. There are two possibilities. First, the set $\{H' : H \xrightarrow{a} H'\}$ is empty. But $G \models \langle a\rangle\mathtt{tt}$ because $G \xrightarrow{a} G'$ and $H \not\models \langle a\rangle\mathtt{tt}$, which contradicts that $G \equiv_{\mathrm{M}} H$. Next, the set $\{H' : H \xrightarrow{a} H'\}$ is non-empty. However it is finite because of image finiteness, and therefore assume it is $\{H_1, \ldots, H_n\}$. Assume $G' \not\equiv_{\mathrm{M}} H_i$ for each $i : 1 \leq i \leq n$, meaning there are formulas $\Phi_1, \ldots, \Phi_n$ such that $G' \models \Phi_i$ and $H_i \not\models \Phi_i$. (Here we use the fact that M is closed under complement; see Section 2.2.) Let Ψ be the formula $\Phi_1 \wedge \ldots \wedge \Phi_n$. Clearly $G' \models \Psi$ and $H_i \not\models \Psi$ for each i, as it fails the ith component of the conjunction. Therefore, $G \models \langle a\rangle\Psi$ because $G \xrightarrow{a} G'$, and $H \not\models \langle a\rangle\Psi$ because each H_i fails to have property Ψ. But this contradicts that $G \equiv_{\mathrm{M}} H$. Therefore, the relation $\equiv_{\mathrm{M}}$ between image-finite processes is a bisimulation. □

Clearly if $E \equiv_\Gamma F$ and $\Delta \subseteq \Gamma$ then $E \equiv_\Delta F$. Therefore, Proposition 1 remains true when Γ is any subset of modal formulas, including the set M. Proposition 2 as illustrated in its proof holds when Γ is also the set M. Under this restriction, these two results are known as the "modal characterization of bisimulation equivalence" due to Hennessy and Milner [29].

Example 1 The need for image finiteness in Proposition 2 is illustrated by the following example. Consider the following family of clocks, $\texttt{Cl}^i$ for $i > 0$,

$$\texttt{Cl}^1 \stackrel{\text{def}}{=} \texttt{tick.0}$$
$$\texttt{Cl}^{i+1} \stackrel{\text{def}}{=} \texttt{tick.Cl}^i \quad i \geq 1$$

and the clock $\texttt{Cl}$ from Section 1.1. Let E be the process $\sum\{\texttt{Cl}^i : i \geq 1\}$, and let F be $E + \texttt{Cl}$. The processes E and F are not bisimilar. The transition $F \stackrel{\texttt{tick}}{\longrightarrow} \texttt{Cl}$ cannot be matched by any transition $E \stackrel{\texttt{tick}}{\longrightarrow} \texttt{Cl}^j$, $j \geq 1$ because $\texttt{Cl} \not\sim \texttt{Cl}^j$. On the other hand $E \equiv_{\mathrm{M}} F$. This follows from the observation that, for any Φ, $\texttt{Cl} \models \Phi$ iff $\exists j \geq 0. \forall k \geq j. \texttt{Cl}^k \models \Phi$ (which is proved later in Section 4.1).

There is an unrestricted characterization of bisimulation equivalence in the case of infinitary modal logic, M_∞, of Section 2.2. The proof is left as an exercise for the reader.

Proposition 3 $E \sim F$ *iff* $E \equiv_{\mathrm{M}_\infty} F$.

A variety of the process equivalences in the linear and branching time spectrum as summarized by Glabbeek in [22] can be presented in terms of having the same modal properties drawn from sublogics of M (when restricted to image-finite processes). Also, these equivalences can often be presented game theoretically by imposing restrictions on possible next moves in a play.

Exercises

1. Recall the definition of α-equivalence from the exercise of the previous section. Consider the restricted case when $\alpha \in \mathbb{N}$. First, is the base case, $E \sim_0 F$ for all E and F. Next, for a successor, $E \sim_{i+1} F$ iff for any $a \in A$,

$$\text{if } E \stackrel{a}{\longrightarrow} E' \text{ then for some } F'.\ F \stackrel{a}{\longrightarrow} F' \text{ and } E' \sim_i F'$$
$$\text{if } F \stackrel{a}{\longrightarrow} F' \text{ then for some } E'.\ E \stackrel{a}{\longrightarrow} E' \text{ and } E' \sim_i F'.$$

Another idea is that of modal depth (defined in the exercises of Section 2.4). The modal depth of an M formula is the maximum embedding of modal operators within it. We let $\mathrm{md}(\Phi)$ be the modal depth of Φ defined as follows:

$$\begin{array}{lclcl} \mathrm{md}(\texttt{tt}) & = & 0 & = & \mathrm{md}(\texttt{ff}) \\ \mathrm{md}(\Phi \wedge \Psi) & = & \max\{\mathrm{md}(\Phi), \mathrm{md}(\Psi)\} & = & \mathrm{md}(\Phi \vee \Psi) \\ \mathrm{md}([K]\Phi) & = & 1 + \mathrm{md}(\Phi) & = & \mathrm{md}(\langle K \rangle \Phi) \end{array}$$

Let M_n be the set of modal M formulas Φ such that $\mathrm{md}(\Phi) \leq n$.

   a. Prove that $E \sim_n F$ iff $E \equiv_{\mathrm{M}_n} F$.

   b. Let E and F be arbitrary image-finite processes, and assume that $E \not\sim F$. Present a method that constructs a formula $\Phi \in \mathrm{M}$ and distinguishing between E and F, that is, for which $E \models \Phi$ and $F \not\models \Phi$.

2. Prove Proposition 3.
3. Give a formula Φ of M_∞ such that $E \models \Phi$ and $F \not\models \Phi$ when E and F are the processes from example 1.
4. Let Γ be the subset of M formulas that do not contain any occurrence of a $[K]$ modality. If E and F are image-finite, prove that $E \equiv_\Gamma F$ iff E and F are simulation equivalent (defined in an exercise of the previous section).
5. A relation B between processes is a 2/3-bisimulation (see Larsen and Skou [37]) provided that, whenever $(E, F) \in B$ and $a \in \mathsf{A}$,

$$\text{if } E \stackrel{a}{\longrightarrow} E' \text{ then for some } F'.\ F \stackrel{a}{\longrightarrow} F' \text{ and } (E', F') \in B$$
$$\text{if } F \stackrel{a}{\longrightarrow} F' \text{ then for some } E'.\ E \stackrel{a}{\longrightarrow} E'.$$

   Let Γ be the subset of M formulas with the restriction that, for any subformula $[K]\Psi$, the formula Ψ is $\mathtt{ff}$. Prove the following for image-finite E and F: $E \equiv_\Gamma F$ iff there are 2/3-bisimulations B and S with $(E, F) \in B$ and $(F, E) \in S$.
6. Let Γ be the subset of M formulas that do not contain an occurrence of a $[K]$ modality within the scope of a $\langle J \rangle$ modality. Provide a definition of an interactive game $\mathsf{G}'(E, F)$ such that player V wins $\mathsf{G}'(E, F)$ iff $E \equiv_\Gamma F$,' assuming that E and F are image-finite.

3.5 Observable bisimulations

Game equivalence and bisimulation equivalence, as we have seen, coincide. Moreover, two equivalent processes have the same modal properties. Conversely, if they are image-finite and have the same M modal properties, then they are bisimilar. There are three different notions here: games, bisimulations and M properties. Not one of this trio abstracts from the silent action τ because each appeals to the family of transition relations $\{\stackrel{a}{\longrightarrow} :\ a \in \mathsf{A}\}$. By consistently replacing this set with the family of observable transitions, as defined in Section 1.3, these notions uniformly abstract from τ. Observable modal logic M^o was defined in Section 2.4 with modalities $[\![K]\!]$, $[\![\ ]\!]$, $\langle\!\langle K \rangle\!\rangle$ and $\langle\!\langle\ \rangle\!\rangle$. Observable games and observable bisimulation relations that appeal to the thicker transition relations $\stackrel{a}{\Longrightarrow}$, $a \in \mathsf{O} \cup \{\varepsilon\}$, are defined below.

A play of the observable game $\mathsf{G}^o(E_0, F_0)$ is a finite or infinite length sequence of pairs $(E_0, F_0) \ldots (E_i, F_i) \ldots$ played by the refuter R and the verifier V. After an initial part of a play $(E_0, F_0) \ldots (E_j, F_j)$, the next pair of processes is determined by one of the following two moves, where $a \in \mathsf{O} \cup \{\varepsilon\}$.

- Player R chooses a transition $E_j \stackrel{a}{\Longrightarrow} E_{j+1}$, then player V chooses a transition with the same label $F_j \stackrel{a}{\Longrightarrow} F_{j+1}$

- Player R chooses a transition $F_j \stackrel{a}{\Longrightarrow} F_{j+1}$, then player V chooses a transition with the same label $E_j \stackrel{a}{\Longrightarrow} E_{j+1}$

The play continues with additional moves.

A position (E_n, F_n) in which one of the processes is able to perform an initial observable action that the other can not is an R-win. A play is won by player R if the play reaches an R-win. Any play that fails to reach such a position counts as a win for player V: this is equivalent to the play having infinite length because player R can always make a move, given that the empty transition $E_j \stackrel{\varepsilon}{\Longrightarrow} E_j$ or $F_j \stackrel{\varepsilon}{\Longrightarrow} F_j$ is available.

As in Section 3.2, a history-free strategy for a player is a set of rules independent of previous moves, and that tell the player how to move. For the refuter, a rule has the form "at position (E, F) choose transition t," where t is either $E \stackrel{a}{\Longrightarrow} E'$ or $F \stackrel{a}{\Longrightarrow} F'$. For the verifier, a rule has the form "at position (E, F) when player R has chosen t choose t'," where t' is corresponding transition of the other process from that of t. A player uses a strategy in a play if all her moves obey the rules in it. A strategy is winning if the player wins every play in which she uses it. For every game $G^o(E, F)$, one of the players has a history-free winning strategy (whose proof is the same as that of Theorem 1 of Section 3.2, except for the use of the thicker transitions). Two processes E and F are "observationally game equivalent" if player V has a winning strategy for $\mathsf{G}^o(E, F)$.

Example 1 The processes $(\mathtt{C} \mid \mathtt{U})\backslash\{\mathtt{in}, \mathtt{ok}\}$ and $\mathtt{Ucop}$ from Section 1.3 are observationally game equivalent. An example play is in Figure 3.6, where the refuter moves from the round vertices and the verifier from the rectangular vertices. The reader is invited to explore other possible plays.

Underpinning observational game equivalence is the existence of an observable bisimulation relation whose definition is as in Section 3.3, except with respect to observable transitions $\stackrel{a}{\Longrightarrow}$.

Definition 1 A binary relation B between processes is an observable bisimulation provided that, whenever $(E, F) \in B$ and $a \in \mathsf{O} \cup \{\varepsilon\}$,

- if $E \stackrel{a}{\Longrightarrow} E'$ then $F \stackrel{a}{\Longrightarrow} F'$ for some F' such that $(E', F') \in B$, and
- if $F \stackrel{a}{\Longrightarrow} F'$ then $E \stackrel{a}{\Longrightarrow} E'$ for some E' such that $(E', F') \in B$

E and F are observably bisimilar, written as $E \approx F$, if there is an observable bisimulation B with $(E, F) \in B$. The relation $\approx$ has many properties in common with $\sim$. It is an equivalence relation. The union of a family of observable bisimulations is also an observable bisimulation (compare Proposition 1 of Section 3.3) and therefore $\approx$ is itself an observable bisimulation. Observable bisimulation equivalence and observable game equivalence also coincide.

Proposition 1 *E is observable game equivalent to F iff $E \approx F$.*

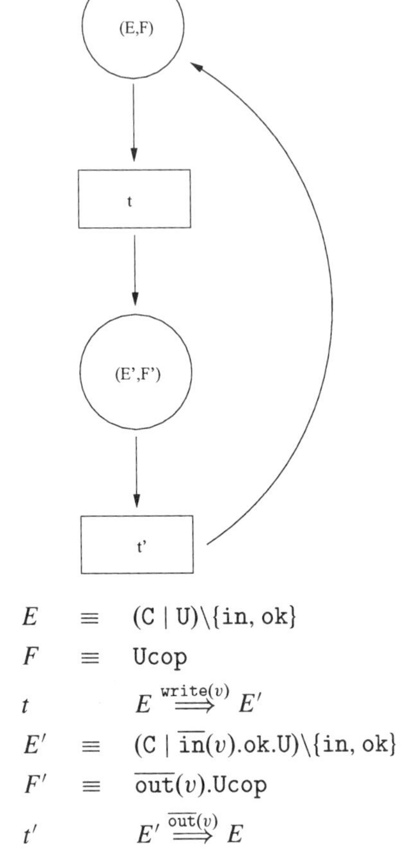

FIGURE 3.6. Game play

The proof of this result is the same as for Proposition 2 of Section 3.3 except, with respect to observable transitions. If processes are bisimilar, then they are also observably bisimilar, as the next result shows.

Proposition 2 *If $E \sim F$, then $E \approx F$.*

Proof. It suffices to show that the relation $\sim$ is an observable bisimulation. The details are left to the reader. □

A direct proof that two processes are observably bisimilar consists of exhibiting an observable bisimulation relation containing them.

Example 2 We show that $\texttt{Protocol} \approx \texttt{Cop}$ by exhibiting an observable bisimulation which contains them. Let B be the following relation.

$$\{(\texttt{Protocol}, \texttt{Cop})\} \cup$$
$$\{((\texttt{Send1}(m) \mid \texttt{Medium} \mid \overline{\texttt{ok}}.\texttt{Receiver})\backslash J,\ \texttt{Cop}) : m \in D\} \cup$$
$$\{((\overline{\texttt{sm}}(m).\texttt{Send1}(m) \mid \texttt{Medium} \mid \texttt{Receiver})\backslash J,\ \overline{\texttt{out}}(m).\texttt{Cop}) : m \in D\} \cup$$
$$\{((\texttt{Send1}(m) \mid \texttt{Med1}(m) \mid \texttt{Receiver})\backslash J,\ \overline{\texttt{out}}(m).\texttt{Cop}) : m \in D\} \cup$$
$$\{((\texttt{Send1}(m) \mid \texttt{Medium} \mid \overline{\texttt{out}}(m).\overline{\texttt{ok}}.\texttt{Receiver})\backslash J,\ \overline{\texttt{out}}(m).\texttt{Cop}) : m \in D\} \cup$$
$$\{((\texttt{Send1}(m) \mid \overline{\texttt{ms}}.\texttt{Medium} \mid \texttt{Receiver})\backslash J,\ \overline{\texttt{out}}(m).\texttt{Cop}) : m \in D\}$$

The reader is invited to establish that B is an observable bisimulation.

There is also an intimate relationship between observable bisimulation equivalence, and having the same properties of observable modal logic M^o of Section 2.4. The following result is the observable correlate of Proposition 1 of Section 3.4. Its proof, which is left as an exercise, is by induction on M^o formulas.

Proposition 3 *If $E \approx F$, then $E \equiv_{\mathrm{M}^o} F$.*

This result is *not* true if we include the modalities $[K]$ and $\langle K \rangle$, or the divergence sensitive modalities of Section 2.5. The converse of Proposition 1 holds for observably image-finite processes. A process E is immediately observably image-finite if, for each $a \in \mathsf{O} \cup \{\varepsilon\}$, the set $\{F : E \overset{a}{\Longrightarrow} F\}$ is finite, and E is observably image-finite if each member of the following set $\{F : \exists w \in (\mathsf{O} \cup \{\varepsilon\})^*.\ E \overset{w}{\Longrightarrow} F\}$ is immediately image-finite.

Proposition 4 *If E and F are observationally image-finite and $E \equiv_{\mathrm{M}^o} F$, then $E \approx F$.*

The proof of this result is very similar to the proof of Proposition 2 of Section 3.4, except that one appeals to observable transitions.

So far there is a smooth passage from results based on transitions $\overset{a}{\longrightarrow}$ to similar results which use observable transitions $\overset{a}{\Longrightarrow}$. There is one important exception, Proposition 3, case 2 of Section 3.3. Observable bisimilarity is not a congruence with respect to the $+$ operator because of the initial preemptive power of τ. The two processes E and $\tau.E$ are observably bisimilar, but for many instances of F the processes $E + F$ and $\tau.E + F$ are not equivalent.

Example 3 The two processes $\tau.2\mathrm{p}.0$ and $2\mathrm{p}.0$ are observably bisimilar, but $\tau.2\mathrm{p}.0 + 1\mathrm{p}.0$ is not equivalent to $2\mathrm{p}.0 + 1\mathrm{p}.0$. For instance, the first process in this pair has the property $\langle\langle\ \rangle\rangle [\![\texttt{1p}]\!] \texttt{ff}$, which the second fails.

In CCS [29, 44], observable equivalence $\approx^c$ is defined as the largest subset of $\approx$ that is also a congruence[10]. Observable bisimilarity is a congruence for all

[10] For instance, $E \approx^c F$ implies $E \approx F$ and for all G, $E + G \approx F + G$.

the operators[11] of CCS, except sum. It therefore turns out that the equivalence $\approx^c$ can be described independently of process contexts in terms of transitions; for it is only the initial preemptive τ transition that causes problems.

Proposition 5 *$E \approx^c F$ iff*

1. *$E \approx F$ and*
2. *if $E \xrightarrow{\tau} E'$, then $F \xrightarrow{\tau} F_1 \stackrel{\varepsilon}{\Longrightarrow} F'$ and $E' \approx F'$ for some F_1 and F', and*
3. *if $F \xrightarrow{\tau} F'$, then $E \xrightarrow{\tau} E_1 \stackrel{\varepsilon}{\Longrightarrow} E'$ and $E' \approx F'$ for some E_1 and E'.*

This Proposition can be viewed as the criterion for when $E \approx^c F$ holds. If E and F are initially unable to perform a silent action (as is the case with `Protocol` and `Cop` of example 2), $E \approx F$ implies $E \approx^c F$.

There is also a finer observable bisimulation equivalence called "branching bisimulation equivalence" which is due to Glabbeek and Weijland [23]. Observable bisimilarity and its congruence are not sensitive to divergence. So, they do not preserve the strong necessity properties discussed in Section 2.5. However it is possible to define equivalences that take divergence into account [26, 31, 61].

Exercises

1. Prove that $\approx$ is an equivalence relation.
2. Let `Cross` be the following simple crossing

   $$\texttt{Cross} \stackrel{\text{def}}{=} \texttt{train}.\overline{\texttt{tcross}}.\texttt{Cross} + \texttt{car}.\overline{\texttt{ccross}}.\texttt{Cross}$$

   and let `Crossing` be as in Figure 1.10. Using games, show that these two crossings are not observably game equivalent.
3. Using games, show that $\texttt{SM}'_n \approx \texttt{SM}_n$ where $\texttt{SM}'_n$ is defined in Section 3.1 and $\texttt{SM}_n$ is as in Figure 1.15.
4. Show that $(\texttt{User} \mid \texttt{Cop})\backslash\{\texttt{in}\}$ is not observably game equivalent to `Ucop`, where `User` and `Cop` are as in Sections 1.1 and 1.2.
5. Prove Propositions 2 and 3 above.
6. Prove the following result. If $E \approx F$, then for any process G, and set of observable actions K, action a, and renaming function f, the following all hold.

   $$\begin{array}{ll} a.E \approx a.F & E \mid G \approx F \mid G \\ E[f] \approx F[f] & E\backslash K \approx F\backslash K \\ E\backslash\backslash K \approx F\backslash\backslash K & E \|_K G \approx F \|_K G \\ Rep(E) \approx Rep(F) & \end{array}$$

7. Prove Proposition 4.
8. Show that if $E \sim F$, then $E \approx^c F$.

---

[11]More generally, only case 2 of Proposition 3 of Section 3.3 fails for $\approx$.

9. Prove Proposition 5

10. Show that $\texttt{Sched}_4 \not\approx \texttt{Sched}'_4$, where these processes are defined in Section 1.4. However, show that

$$\texttt{Sched}_4\backslash\backslash\{b_1, \ldots, b_4\} \approx \texttt{Sched}'_4\backslash\backslash\{b_1, \ldots, b_4\}.$$

11. For $a \in \mathsf{A}$ let $\hat{a}$ be a if $a \neq \tau$, and let $\hat{\tau}$ be ε. A binary relation B between processes is an ob bisimulation just in case whenever $(E, F) \in B$ and $a \in \mathsf{A}$,

    a. if $E \xrightarrow{a} E'$ then $F \stackrel{\hat{a}}{\Longrightarrow} F'$ for some F' such that $(E', F') \in B$, and

    b. if $F \xrightarrow{a} F'$ then $E \stackrel{\hat{a}}{\Longrightarrow} E'$ for some E' such that $(E', F') \in B$.

    Two processes are ob equivalent, denoted by $\approx'$, if they are related by an ob bisimulation relation. Prove that $\approx = \approx'$.

12. Prove that $\texttt{Ct}_0 \approx^c \texttt{Count}$, where these processes are given in Figure 1.4 and Section 1.6.

13. Extend the modal logic M^o so that two image-finite processes have the same modal properties if, and only if, they are observably congruent.

3.6 Equivalence checking

A direct proof that two processes are bisimilar, or observably equivalent, is to exhibit the appropriate bisimulation relation that contains them. Examples in Sections 3.3 and 3.5 show the proof technique. In the case that processes are finite state, this can be done automatically. There is a variety of tools that include this capability including the Edinburgh Concurrency Workbench [14].

Alternatively, equivalence proofs can utilize conditional equational reasoning. There is an assortment of algebraic, and semi-algebraic, theories of processes depending on the equivalence and the process combinators. For details, see the references [26, 31, 3, 44]. It is essential that the equivalence be a congruence. To give a flavour of equational reasoning, we present a proof in the equational theory for CCS that a simplified slot machine without data values is equivalent to a streamlined process description. The equivalence involved is $\approx^c$, the observational congruence defined in the previous section.

The following important CCS laws are used in the proof. The variables x and y stand for arbitrary CCS process expressions.

$$
\begin{array}{lcll}
a.\tau.x & = & a.x & \\
x + \tau.x & = & \tau.x & \\
(x+y)\backslash K & = & x\backslash K + y\backslash K & \\
(a.x)\backslash K & = & a.(x\backslash K) & \text{if } a \notin K \cup \overline{K} \\
(a.x)\backslash K & = & 0 & \text{if } a \in K \cup \overline{K} \\
x + 0 & = & x &
\end{array}
$$

The last four are clear from the behavioural meanings of the operators. The first two are τ-laws and show that we are dealing with an observable equivalence. We shall also appeal to a rule schema called an "expansion law" by Milner [44], relating concurrency and choice.

$$\begin{array}{ll}
\text{if} & x_i = \sum\{a_{ij}.x_{ij} \,:\, 1 \le j \le n_i\} \text{ for } i : 1 \le i \le m, \\
\text{then} & x_1 \mid \ldots \mid x_m \;=\; \sum\{a_{ij}.y_{ij} \,:\, 1 \le i \le m \text{ and } 1 \le j \le n_i\} \\
 & \qquad + \sum\{\tau.y_{klij} \,:\, 1 \le k < i \le m \text{ and } a_{kl} = \overline{a}_{ij}\}, \\
\text{where} & y_{ij} \equiv x_1 \mid \ldots \mid x_{i-1} \mid x_{ij} \mid x_{i+1} \mid \ldots \mid x_m \\
\text{and} & y_{klij} \equiv x_1 \mid \ldots \mid x_{k-1} \mid x_{kl} \mid x_{k+1} \mid \ldots \mid x_{i-1} \mid x_{ij} \mid x_{i+1} \mid \ldots \mid x_m.
\end{array}$$

For example, if

$$\begin{array}{rcl}
x_1 & = & a.x_{11} + b.x_{12} + a.x_{13} \\
x_2 & = & \overline{a}.x_{21} + c.x_{22},
\end{array}$$

then

$$\begin{array}{rcl}
x_1 | x_2 & = & a.(x_{11}|x_2) + b.(x_{12}|x_2) + a.(x_{13}|x_2) + \overline{a}.(x_1|x_{21}) + \\
 & & c.(x_1|x_{22}) + \tau.(x_{11}|x_{21}) + \tau.(x_{13}|x_{21}).
\end{array}$$

The expansion rule is justified by the transition rules for the parallel operator.

Proof rules for recursion ($\stackrel{\text{def}}{=}$) are also needed. If E does not contain any occurrences of the parallel operator $|$, then P is said to be "guarded" in E, provided that all occurrences of P in E are within the scope of a prefix $a.$ and a is an observable action (that is, not τ). Assume that P is the only process constant in E. The guardedness condition guarantees that the equation $P = E$ has a unique solution up to $\approx^c$. A solution to the equation $P = E$ is a process F such that $F \approx^c E\{F/P\}$. Uniqueness of solution is that, if both F and G are solutions, then $F \approx^c G$.

Example 1 The clock $\texttt{Cl}$ is a solution to the equation $P = \texttt{tick}.P$ because $\texttt{Cl} \approx^c \texttt{tick}.\texttt{Cl}$. Moreover, any other solution E (such as $\texttt{Cl}_2$) has the property that $\texttt{Cl} \approx^c E$. In contrast, the equation $P = \tau.P$ where P is not guarded has as solutions any process $\tau.E$ because $\tau.E \approx^c \tau.\tau.E$.

The specific recursion proof rules used are the following.

- if $P \stackrel{\text{def}}{=} E$ then $P = E$
- if $P = E$ and P is guarded in E, and $Q = F$ and Q is guarded in F, and $E\{Q/P\} = F$, then $P = F$

$$\begin{array}{lcl}
\mathtt{IO} & \stackrel{\mathrm{def}}{=} & \mathtt{slot}.\mathtt{IO}_1 \\
\mathtt{IO}_1 & \stackrel{\mathrm{def}}{=} & \overline{\mathtt{bank}}.\mathtt{IO}_2 \\
\mathtt{IO}_2 & \stackrel{\mathrm{def}}{=} & \mathtt{lost}.\overline{\mathtt{loss}}.\mathtt{IO} + \mathtt{release}.\overline{\mathtt{win}}.\mathtt{IO}
\end{array}$$

$$\begin{array}{lcl}
\mathtt{B} & \stackrel{\mathrm{def}}{=} & \mathtt{bank}.\mathtt{B}_1 \\
\mathtt{B}_1 & \stackrel{\mathrm{def}}{=} & \overline{\mathtt{max}}.\mathtt{left}.\mathtt{B}
\end{array}$$

$$\begin{array}{lcl}
\mathtt{D} & \stackrel{\mathrm{def}}{=} & \mathtt{max}.\mathtt{D}_1 \\
\mathtt{D}_1 & \stackrel{\mathrm{def}}{=} & \overline{\mathtt{lost}}.\overline{\mathtt{left}}.\mathtt{D} + \overline{\mathtt{release}}.\overline{\mathtt{left}}.\mathtt{D}
\end{array}$$

$$\mathtt{SM} \equiv (\mathtt{IO} \mid \mathtt{B} \mid \mathtt{D})\backslash K \text{ where } K = \{\mathtt{bank}, \mathtt{max}, \mathtt{left}, \mathtt{lost}, \mathtt{release}\}$$

$$\mathtt{SM}' \stackrel{\mathrm{def}}{=} \mathtt{slot}.(\tau.\overline{\mathtt{loss}}.\mathtt{SM}' + \tau.\overline{\mathtt{win}}.\mathtt{SM}')$$

FIGURE 3.7. A simplified slot machine

Example 2 The recursion rules can be used to prove that $\mathtt{Cl} \approx^c \mathtt{Cl}_2$ (where these processes are pictured in Figure 3.1) as follows.

$\mathtt{Cl} = \mathtt{tick}.\mathtt{Cl}$	by definition of $\mathtt{Cl}$
$\mathtt{tick}.\mathtt{Cl} = \mathtt{tick}.\mathtt{tick}.\mathtt{Cl}$	by congruence
$\mathtt{Cl} = \mathtt{tick}.\mathtt{tick}.\mathtt{Cl}$	by transitivity of $=$
$\mathtt{Cl}_2 = \mathtt{tick}.\mathtt{tick}.\mathtt{Cl}_2$	by definition of $\mathtt{Cl}_2$
$(\mathtt{tick}.\mathtt{tick}.\mathtt{Cl})\{\mathtt{Cl}_2/\mathtt{Cl}\} = \mathtt{tick}.\mathtt{tick}.\mathtt{Cl}_2$	by equality
$\mathtt{Cl} = \mathtt{Cl}_2$	by the second recursion rule

The slot machine SM without data values, and its succinct description SM′, appear in Figure 3.7. We prove that $\mathtt{SM} = \mathtt{SM}'$. The idea behind the proof is to first simplify SM by showing that it is equal to an expression E that does not contain the parallel operator. The proof proceeds on SM using the expansion law, and the laws earlier for $\backslash K$, 0 and τ (and the first recursion rule).

$$\begin{array}{lcl}
\mathtt{SM} & = & (\mathtt{IO} \mid \mathtt{B} \mid \mathtt{D})\backslash K \\
& = & (\mathtt{slot}.\mathtt{IO}_1 \mid \mathtt{bank}.\mathtt{B}_1 \mid \mathtt{max}.\mathtt{D}_1)\backslash K \\
& = & (\mathtt{slot}.(\mathtt{IO}_1 \mid \mathtt{B} \mid \mathtt{D}) + \mathtt{bank}.(\mathtt{IO} \mid \mathtt{B}_1 \mid \mathtt{D}) + \mathtt{max}.(\mathtt{IO} \mid \mathtt{B} \mid \mathtt{D}_1))\backslash K \\
& = & (\mathtt{slot}.(\mathtt{IO}_1 \mid \mathtt{B} \mid \mathtt{D}))\backslash K + (\mathtt{bank}.(\mathtt{IO} \mid \mathtt{B}_1 \mid \mathtt{D}))\backslash K + (\mathtt{max}.(\mathtt{IO} \mid \mathtt{B} \mid \mathtt{D}_1))\backslash K \\
& = & \mathtt{slot}.(\mathtt{IO}_1 \mid \mathtt{B} \mid \mathtt{D})\backslash K + 0 + 0 \\
& = & \mathtt{slot}.(\mathtt{IO}_1 \mid \mathtt{B} \mid \mathtt{D})\backslash K
\end{array}$$

Let $\mathtt{SM}_1 \equiv (\mathtt{IO}_1 \mid \mathtt{B} \mid \mathtt{D})\backslash K$. By similar reasoning to the above, we obtain

$$\mathtt{SM}_1 = \tau.\tau.(\mathtt{IO}_2 \mid \mathtt{left.B} \mid \mathtt{D}_1)\backslash K.$$

Assume $\mathtt{SM}_2 \equiv (\mathtt{IO}_2 \mid \mathtt{left.B} \mid \mathtt{D}_1)\backslash K$. By similar reasoning,

$$\begin{aligned}
\mathtt{SM}_2 &= \tau.\mathtt{SM}_3 + \tau.\mathtt{SM}_4 \quad \text{where} \\
\mathtt{SM}_3 &\equiv (\overline{\mathtt{loss}}.\mathtt{IO} \mid \mathtt{left.B} \mid \overline{\mathtt{left}}.\mathtt{D})\backslash K \\
\mathtt{SM}_4 &\equiv (\overline{\mathtt{win}}.\mathtt{IO} \mid \mathtt{left.B} \mid \overline{\mathtt{left}}.\mathtt{D})\backslash K.
\end{aligned}$$

The τ-laws are used in the following chain of reasoning.

$$\begin{aligned}
\mathtt{SM}_3 &= (\overline{\mathtt{loss}}.\mathtt{IO} \mid \mathtt{left.B} \mid \overline{\mathtt{left}}.\mathtt{D})\backslash K \\
&= \overline{\mathtt{loss}}.(\mathtt{IO} \mid \mathtt{left.B} \mid \overline{\mathtt{left}}.\mathtt{D})\backslash K + \tau.(\overline{\mathtt{loss}}.\mathtt{IO} \mid \mathtt{B} \mid \mathtt{D})\backslash K \\
&= \overline{\mathtt{loss}}.(\tau.\mathtt{SM} + \mathtt{slot}.(\mathtt{IO}_1 \mid \mathtt{left.B} \mid \overline{\mathtt{left}}.\mathtt{D})\backslash K) + \tau.\overline{\mathtt{loss}}.\mathtt{SM} \\
&= \overline{\mathtt{loss}}.(\tau.\mathtt{SM} + \mathtt{slot}.\tau.(\mathtt{IO}_1 \mid \mathtt{B} \mid \mathtt{D})\backslash K) + \tau.\overline{\mathtt{loss}}.\mathtt{SM} \\
&= \overline{\mathtt{loss}}.(\tau.\mathtt{SM} + \mathtt{slot}.(\mathtt{IO}_1 \mid \mathtt{B} \mid \mathtt{D})\backslash K) + \tau.\overline{\mathtt{loss}}.\mathtt{SM} \\
&= \overline{\mathtt{loss}}.(\tau.\mathtt{SM} + \mathtt{SM}) + \tau.\overline{\mathtt{loss}}.\mathtt{SM} \\
&= \overline{\mathtt{loss}}.\tau.\mathtt{SM} + \tau.\overline{\mathtt{loss}}.\mathtt{SM} \\
&= \overline{\mathtt{loss}}.\mathtt{SM} + \tau.\overline{\mathtt{loss}}.\mathtt{SM} \\
&= \tau.\overline{\mathtt{loss}}.\mathtt{SM}
\end{aligned}$$

By similar reasoning, $\mathtt{SM}_4 = \tau.\overline{\mathtt{win}}.\mathtt{SM}$. Backtracking and substituting equals for equals, and then applying τ laws gives the following.

$$\begin{aligned}
\mathtt{SM} &= \mathtt{slot}.\mathtt{SM}_1 \\
&= \mathtt{slot}.\tau.\tau.\mathtt{SM}_2 \\
&= \mathtt{slot}.\tau.\tau.(\tau.\mathtt{SM}_3 + \tau.\mathtt{SM}_4) \\
&= \mathtt{slot}.\tau.\tau.(\tau.\tau.\overline{\mathtt{loss}}.\mathtt{SM} + \tau.\tau.\overline{\mathtt{win}}.\mathtt{SM}) \\
&= \mathtt{slot}.(\tau.\overline{\mathtt{loss}}.\mathtt{SM} + \tau.\overline{\mathtt{win}}.\mathtt{SM})
\end{aligned}$$

We have now shown that $\mathtt{SM} = E$, where E does not contain the parallel operator, and where $\mathtt{SM}$ is guarded in E. The expression E is very close to the definition of $\mathtt{SM}'$ (and $\mathtt{SM}'$ is also guarded within it). Clearly, $E\{\mathtt{SM}'/\mathtt{SM}\} = \mathtt{slot}.(\tau.\overline{\mathtt{loss}}.\mathtt{SM}' + \tau.\overline{\mathtt{win}}.\mathtt{SM}')$, so by the second recursion rule $\mathtt{SM} = \mathtt{SM}'$, which completes the proof.

Exercises

1. For each of the following cases of x_1, x_2 and x_3, define $x_1 \mid x_2 \mid x_3$ using the expansion theorem.

   a.

$$\begin{aligned}
x_1 &= a.x_{11} + b.x_{12} + a.x_{12} \\
x_2 &= \overline{a}.x_{21} + c.x_{22} \\
x_3 &= \overline{a}.x_{31} + \overline{c}.x_{32}
\end{aligned}$$

b.

$$x_1 = \tau.x_{11} + \tau.x_{12}$$
$$x_2 = a.x_{21} + b.x_{22}$$
$$x_3 = \overline{a}.x_{31} + \overline{a}.x_{32} + \tau.x_{33}$$

c.

$$x_1 = a.x_{11} + b.x_{12} + a.x_{12} + \overline{a}.x_{13}$$
$$x_2 = \overline{a}.x_{21} + \overline{b}.x_{22} + a.x_{23}$$
$$x_3 = \overline{a}.x_{31} + b.x_{32} + a.x_{33}$$

2. Refine the expansion law to take account of the restriction operator. That is, assume that each x_i has the form $\sum\{a_{ij}.x_{ij} \; : \; 1 \le j \le n_i\}$ and the parallel form is $(x_1 \mid \ldots \mid x_m)\backslash K$.

3. Let $\texttt{Cl}' \stackrel{\text{def}}{=} \texttt{tick.tick.tick.Cl}'$. Use the recursion proof rules to prove that $\texttt{Cl}' = \texttt{Cl}_2$.

4. Assume that there is just one datum value in the set D. Using equational reasoning only, prove that $\texttt{Protocol} \approx^c \texttt{Cop}$ (where these processes are defined in Chapter 1).

5. Compare the different methods for proving equivalence, by showing that $\texttt{SM} \approx \texttt{SM}'$:

- **a.** directly using games
- **b.** directly by exhibiting an observable bisimulation
- **c.** directly by showing that they obey the same modal properties in $\mathcal{M}^o$

6. Extend the proof rules of the equational system to take into account value passing, and then prove equationally that $\texttt{SM}_\texttt{n} = \texttt{SM}'_\texttt{n}$.

7. Extend the second recursion rule so that the expressions E and F can contain occurrences of parallel. (To do this we need to refine the definition of being guarded.)

4

Temporal Properties

Modal logics as introduced in Chapter 2 can express local capabilities and necessities of processes such as "`tick` is a possible next action" or "`tick` must happen next." However, they cannot express enduring capabilities such as "`tick` is *always* a possible next action" or long term inevitabilities such as "`tick` *eventually* happens." These features, especially in the guise of safety or liveness properties, have been found to be very useful when analysing the behaviour of concurrent systems. Another abstraction from behaviour is a run of a process that is a finite or infinite length sequence of transitions. Runs provide a basis for understanding longer term capabilities. Logics where properties are primarily ascribed to runs of systems are called "temporal" logics. An alternative foundation for temporal logic is to view enduring features as extremal solutions to recursive modal equations.

4.1 Modal properties revisited

A property partitions a family of processes into two disjoint sets, the subset of processes that have the property and the subset that does not have the property. For example, the formula $\langle \texttt{tick} \rangle \texttt{tt}$ divides $\{\texttt{Cl}_1, \texttt{tock.Cl}_1\}$ into two subsets $\{\texttt{Cl}_1\}$ and $\{\texttt{tock.Cl}_1\}$. Given a formula Φ and a set of processes E, we define $\| \Phi \|^{\mathsf{E}}$ be the subset of processes in E having the modal property Φ.

$$\| \Phi \|^{\mathsf{E}} \stackrel{\text{def}}{=} \{E \in \mathsf{E} : E \models \Phi\}$$

Therefore, each modal formula Φ partitions E into $\| \Phi \|^{\mathsf{E}}$ and $\mathsf{E} - \| \Phi \|^{\mathsf{E}}$.

There are many different notations for describing properties of processes. Modal logic is one such formalism. Other notations are also logical and include first-order and second-order logic over transition graphs. Another kind of formalism is automata, which recognises words, trees or graphs. The expressive power of different notations can be compared by examining the properties of processes they are able to define. Whatever the notation, a property of a set of processes can be identified with the subset that has it. Consider the following family of clocks $\mathsf{E} = \{\texttt{Cl}^i, \texttt{Cl} : i \geq 1\}$, where $\texttt{Cl}^i$ is defined in example 1 of Section 3.4. For instance, the transition graph for $\texttt{Cl}_4$ is as follows.

$$\texttt{Cl}^4 \stackrel{\texttt{tick}}{\longrightarrow} \texttt{Cl}^3 \stackrel{\texttt{tick}}{\longrightarrow} \texttt{Cl}^2 \stackrel{\texttt{tick}}{\longrightarrow} \texttt{Cl}^1 \stackrel{\texttt{tick}}{\longrightarrow} 0$$

$\texttt{Cl}$ is distinguishable from other members of E because of its long term capability for ticking endlessly. Each clock $\texttt{Cl}^i$ ticks exactly i times before stopping. The property "can tick forever" partitions E into two subsets $\{\texttt{Cl}\}$ and $\mathsf{E} - \{\texttt{Cl}\}$. However, this partition cannot be captured by a single modal formula, as the following result shows.

Proposition 1 *For any modal $\Phi \in \mathrm{M} \cup \mathrm{M}^{o\downarrow}$, if $\texttt{Cl} \models \Phi$, then there is a $j \geq 1$ such that, for all $k \geq j$, $\texttt{Cl}^k \models \Phi$.*

Proof. By induction on the structure of Φ. The base case when Φ is $\texttt{tt}$ or $\texttt{ff}$ is clear. The induction step divides into subcases. Let $\Phi = \Psi_1 \wedge \Psi_2$ and assume that $\texttt{Cl} \models \Phi$. Therefore, $\texttt{Cl} \models \Psi_1$ and $\texttt{Cl} \models \Psi_2$. By the induction hypothesis, there is a $j1 \geq 1$ and a $j2 \geq 1$ such that $\texttt{Cl}^{k1} \models \Psi_1$ for all $k1 \geq j1$ and $\texttt{Cl}^{k2} \models \Psi_2$ for all $k2 \geq j2$. Let j be the maximum of $\{j1, j2\}$, and so $\texttt{Cl}^k \models \Phi$ for all $k \geq j$. Next, assume that $\Phi = \Psi_1 \vee \Psi_2$ and $\texttt{Cl} \models \Phi$. So $\texttt{Cl} \models \Psi_1$ or $\texttt{Cl} \models \Psi_2$. Suppose it is the first of these two. By the induction hypothesis, there is a $j \geq 1$ such that $\texttt{Cl}^k \models \Psi_1$ for all $k \geq j$, and therefore also $\texttt{Cl}^k \models \Phi$ for each $k \geq j$. Let $\Phi = [K]\Psi$ and assume that $\texttt{Cl} \models \Phi$. If $\texttt{tick} \notin K$, then $\texttt{Cl}^i \models \Phi$ for all $i \geq 1$. Otherwise $\texttt{tick} \in K$, and since $\texttt{Cl} \stackrel{\texttt{tick}}{\longrightarrow} \texttt{Cl}$ it follows that $\texttt{Cl} \models \Psi$. By the induction hypothesis, there is a $j \geq 1$ such that $\texttt{Cl}^k \models \Psi$ for all $k \geq j$. However, since $\texttt{Cl}^{i+1} \stackrel{\texttt{tick}}{\longrightarrow} \texttt{Cl}^i$ it follows that $\texttt{Cl}^k \models \Phi$ for all $k \geq j+1$. The case $\Phi = \langle K \rangle \Psi$ is similar. The other modal cases when Φ is $\langle\!\langle\ \rangle\!\rangle \Psi$, $[\![\ ]\!] \Psi$, $[\![\downarrow]\!] \Psi$, or $\langle\!\langle \uparrow \rangle\!\rangle \Psi$ are more straightforward and are left as an exercise. □

Proposition 1 shows that no modal formula partitions the set E into the pair $\{\texttt{Cl}\}$ and $\mathsf{E} - \{\texttt{Cl}\}$, and therefore the enduring capability of being able to tick forever is *not* expressible in the modal logics introduced in Chapter 2. A similar argument establishes that the property "`tick` eventually happens" is also not definable within these modal logics.

Exercises

1. **a.** Define $\| \Phi \|^{\mathsf{E}}$ directly by induction on the structure of Φ. What assumptions do you make about the set E?

   b. Let E be the the set $\{\mathtt{Ct}_i \, : \, i \geq 0\}$ of counters from Figure 1.4. Work out the following sets using your inductive definition.

   i. $\| \langle \mathtt{down} \rangle \mathtt{tt} \wedge \langle \mathtt{up} \rangle \mathtt{tt} \|^{\mathsf{E}}$

   ii. $\| [\mathtt{down}](\langle \mathtt{down} \rangle \mathtt{tt} \wedge \langle \mathtt{round} \rangle \mathtt{tt}) \|^{\mathsf{E}}$

   iii. $\| [\mathtt{up}]\mathtt{ff} \|^{\mathsf{E}}$

   c. Assume instead that E is the subset $\{\mathtt{Ct}_{2i} \, : \, i \geq 0\}$. What sets are now determined by the formulas above, according to your inductive definition?

2. An important feature of modal logic is that its basic modal operators are monotonic with respect to subset inclusion. Prove the following, assuming that $\# \in \{[K], \langle K \rangle, [\![\ ]\!], \langle\!\langle\ \rangle\!\rangle, [\![\downarrow]\!], \langle\!\langle \uparrow \rangle\!\rangle\}$ in part c.

   a. If $\mathsf{E}_1 \subseteq \mathsf{E}_2$ then $\| \Phi \|^{\mathsf{E}} \cap \mathsf{E}_1 \subseteq \| \Phi \|^{\mathsf{E}} \cap \mathsf{E}_2$

   b. If $\mathsf{E}_1 \subseteq \mathsf{E}_2$ then $\| \Phi \|^{\mathsf{E}} \cup \mathsf{E}_1 \subseteq \| \Phi \|^{\mathsf{E}} \cup \mathsf{E}_2$

   c. If $\| \Phi \|^{\mathsf{E}} \subseteq \| \Psi \|^{\mathsf{E}}$ then $\| \# \Phi \|^{\mathsf{E}} \subseteq \| \# \Psi \|^{\mathsf{E}}$

3. We can view the meaning of a modal operator $\#$ as a process transformer $\| \# \|^{\mathsf{E}}$ mapping subsets of E to subsets of E. For example, $\| [K] \|^{\mathsf{E}}$ is the function which for any $\mathsf{E}_1 \subseteq \mathsf{E}$ is defined as follows.

   $$\| [K] \|^{\mathsf{E}} \mathsf{E}_1 \; = \; \{F \in \mathsf{E} \, : \, \text{if } F \stackrel{a}{\longrightarrow} E \text{ and } a \in K \text{ then } E \in \mathsf{E}_1\}$$

   a. Define the transformers $\| \langle K \rangle \|^{\mathsf{E}}$, $\| \langle\!\langle K \rangle\!\rangle \|^{\mathsf{E}}$ and $\| [\![\downarrow]\!] \|^{\mathsf{E}}$.

   b. An indexed family of subsets $\{\mathsf{E}_i \subseteq \mathsf{E} \, : \, i \geq 0\}$ of E is a chain if $\mathsf{E}_i \subseteq \mathsf{E}_j$ when $i \leq j$. A modal operator $\#$ is $\cup$-continuous if for any such chain the set $\| \# \|^{\mathsf{E}} \bigcup \{\mathsf{E}_i \, : \, i \geq 0\}$ is the same as $\bigcup \{\| \# \|^{\mathsf{E}} \mathsf{E}_i \, : \, i \geq 0\}$. Prove that, if each member of E is finitely branching (that is $\{F \, : \, E \stackrel{a}{\longrightarrow} F \text{ with } a \in \mathsf{A}\}$ is finite for each $E \in \mathsf{E}$), then $[K]$ is $\cup$-continuous. Give an example process which fails $\cup$-continuity.

4. Prove that the property "`tick` eventually happens" is not definable within $\mathrm{M} \cup \mathrm{M}^{o\downarrow}$.

4.2 Processes and their runs

Proposition 1 of the previous section shows that modal logic is not very expressive. Although able to describe local or immediate capabilities and necessities, modal formulas cannot capture global or long term features of processes. Consider the contrast between the local capability for ticking and the enduring capability for ticking forever, and the contrast between the urgent inevitability that tick must happen next with the lingering inevitability that tick eventually happens.

Another abstraction from behaviour is a run of a process. A run of E_0 is a finite or infinite length sequence of transitions

$$E_0 \xrightarrow{a_1} E_1 \xrightarrow{a_2} \dots \xrightarrow{a_n} E_n \xrightarrow{a_{n+1}} \dots$$

with "maximal" length. This means that, if a run has finite length, then its final process is unable to perform a transition because, otherwise, the sequence can be extended. A deadlocked process E only has the zero length run E. A run of a process carves out a path through its transition graph.

Example 1 $\mathtt{Cl}_1 \xrightarrow{\mathtt{tick}} \mathtt{tock.Cl}_1 \xrightarrow{\mathtt{tock}} \mathtt{Cl}_1 \xrightarrow{\mathtt{tick}} \dots$ is the only run from the clock $\mathtt{Cl}_1$ and it has infinite length. $\mathtt{Cl}_5$[1] has infinitely many finite length runs, each of the form $\mathtt{Cl}_5 \xrightarrow{\mathtt{tick}} \dots \xrightarrow{\mathtt{tick}} \mathtt{Cl}_5 \xrightarrow{\mathtt{tock}} \mathtt{0}$ and the single infinite length run $\mathtt{Cl}_5 \xrightarrow{\mathtt{tick}} \mathtt{Cl}_5 \xrightarrow{\mathtt{tick}} \dots$. An infinite length cyclic run of Crossing is

$$\mathtt{Crossing} \xrightarrow{\mathtt{car}} E_1 \xrightarrow{\tau} E_4 \xrightarrow{\overline{\mathtt{ccross}}} E_8 \xrightarrow{\tau} \mathtt{Crossing} \xrightarrow{\mathtt{car}} \dots,$$

where E_1, E_4, and E_8 are as in Figure 1.12.

Runs provide a means for distinguishing between local and long term features of processes. For instance, a process has the capability for ticking forever provided that it has a perpetual run of tick transitions. The property of a process that tick eventually happens is the requirement that, within every run of it, there is at least one tick transition. The clock $\mathtt{Cl}_1$ of Example 1 has the property that tock eventually happens. However, the other clock $\mathtt{Cl}_5$ fails to have this trait because of its sole infinite length run, which does not contain the action tock.

Bisimulation equivalence "preserves" runs in the sense that, if two processes are bisimilar then for any run of one of the processes there is a corresponding run of the other process.

Proposition 1 *Assume that* $E_0 \sim F_0$.

1. *If* $E_0 \xrightarrow{a_1} E_1 \xrightarrow{a_2} \dots \xrightarrow{a_n} E_n$ *is a finite length run, then there is a run* $F_0 \xrightarrow{a_1} F_1 \xrightarrow{a_2} \dots \xrightarrow{a_n} F_n$ *such that* $E_i \sim F_i$ *for all* $i : 0 \leq i \leq n$; *and*
2. *If* $E_0 \xrightarrow{a_1} E_1 \xrightarrow{a_2} \dots$ *is an infinite length run, then there is an infinite length run* $F_0 \xrightarrow{a_1} F_1 \xrightarrow{a_2} \dots$ *such that* $E_i \sim F_i$ *for all* $i \geq 0$.

Because bisimulation equivalence is symmetric[2], Proposition 1 also implies that every run from F_0 has to be matched with a corresponding run from E_0. We leave the proof of this result as an exercise for the reader.

---

[1] $\mathtt{Cl}_5 \stackrel{\text{def}}{=} \mathtt{tick.Cl}_5 + \mathtt{tock.0}$.

[2] If $E_0 \sim F_0$, then also $F_0 \sim E_0$.

Many significant properties of systems can be understood as features of their runs. Especially important is a classification of properties into "safety" and "liveness," originally due to Lamport [35]. A safety property states "nothing bad ever happens," whereas a liveness property expresses "something good eventually happens." A process has a safety property if none of its runs has the bad feature, and it has a liveness property if all of its runs have the good feature. That is, we can informally express them as follows.

Safety(Φ)	for all runs π, Φ is never true in π
Liveness(Φ)	for all runs π, Φ is true in π

Example 2 A property that distinguishes each clock $\mathtt{Cl}^i$ of the previous section from `Cl` is eventual termination. The good feature is expressed by the formula $[-]\mathtt{ff}$, and the property is given as Liveness($[-]\mathtt{ff}$). On the other hand, termination can also be viewed as defective, as exhaustion of the clock. In this case, `Cl` has the safety property of absence of deadlock, expressed as Safety($\langle - \rangle \mathtt{tt}$), which each $\mathtt{Cl}^i$ fails to have.

Liveness and safety properties of a process pertain to all of its runs. Weaker properties relate to some runs of a process. A "weak" safety property states "in some run nothing bad ever happens," and a "weak" liveness property asserts "in some run something good eventually happens."

WSafety(Φ)	for some run π, Φ is never true in π
WLiveness(Φ)	for some run π, Φ is true in π

Notice that a weak safety property is the "dual" of a liveness property (and a weak liveness property is the "dual" of a safety property).

WSafety(Φ)	iff	not(Liveness(notΦ))
WLiveness(Φ)	iff	not(Safety(notΦ))

Example 3 A weak liveness property of $\mathtt{SM}_n$ of Figure 1.15 is that it may eventually pay out a windfall of a million pounds. The good feature is $\langle \overline{\mathtt{win}}(10^6) \rangle \mathtt{tt}$, and so the property is given by WLiveness($\langle \overline{\mathtt{win}}(10^6) \rangle \mathtt{tt}$).

There are also intermediate cases between all and some runs when liveness or safety properties pertain to special families of runs which obey some general constraints.

Example 4 A desirable property of `Crossing` is that, whenever a car approaches, eventually it crosses. This requirement is that any run containing the action `car` also contains $\overline{\mathtt{ccross}}$ as a later action.

Sometimes, the relevant constraints are complex and depend on assumptions outwith the possible behaviour of the process itself. For example, `Protocol` of Figure 1.13 fails to have the property "whenever a message is input eventually it is output" because of runs where a message is forever retransmitted. However, we may assume that the medium does eventually pass to the receiver a repeatedly retransmitted message, meaning these deficient runs are thereby precluded.

Exercises

1. Enumerate the runs of `Ven` and `Crossing`.
2. Prove Proposition 1.
3. For each of the following, state whether it is a liveness or a safety property, and identify the good or bad feature.
   a. At most five messages are in the buffer at any time.
   b. The cost of living never decreases.
   c. The temperature never rises.
   d. All good things must come to an end.
   e. If an interrupt occurs, then a message is printed within one second.
4. Prove the following dualities.

$$\begin{array}{lll} \text{WSafety}(\Phi) & \text{iff} & \text{not}(\text{Liveness}(\text{not}\Phi)) \\ \text{WLiveness}(\Phi) & \text{iff} & \text{not}(\text{Safety}(\text{not}\Phi)) \end{array}$$

5. Observable bisimulation equivalence, $\approx$, does not "preserve" runs in the sense of Proposition 1. For instance, $\tau.0$ does not have a corresponding infinite length run to the following run of Div (where $\mathtt{Div} \stackrel{\mathrm{def}}{=} \tau.\mathtt{Div}$)

$$\mathtt{Div} \xrightarrow{\tau} \mathtt{Div} \xrightarrow{\tau} \ldots$$

   even though $\mathtt{Div} \approx \tau.0$. We may try to weaken the matching requirement by stipulating that, for any run from one process, there is a corresponding run from the other process such that there is a finite or an infinite partition across these runs containing equivalent processes.
   a. Spell out a definition of equivalence $\equiv$ based on this partitioning requirement.
   b. Prove $a.(E + \tau.F) + a.E \approx a.(E + \tau.F)$ for any E and F. Is this still true if you replace $\approx$ with $\equiv$?
   c. What is the relation between $\equiv$ and branching bisimulation, as defined by Glabbeek and Weijland [23]?
6. Prove that the following properties are not definable within modal logic $\mathrm{M} \cup \mathrm{M}^{o\downarrow}$.
   a. "in some run `tick` does not happen"
   b. "in some run eventually `tick` happens"
   c. "the sequence of actions $a_1 \ldots a_4$ happens cyclically forever starting with a_1"

4.3 The temporal logic CTL

Modal logic expresses properties of processes in terms of their transitions. Temporal logic, on the other hand, ascribes properties to processes by expressing features of their runs. In fact, there is not a clear demarcation between modal and temporal logic because modal operators can also be viewed as temporal operators as follows.

$$E_0 \models [K]\Phi \quad \text{iff} \quad \text{for any run } E_0 \xrightarrow{a_1} E_1 \dots \text{ if } a_1 \in K \text{ then } E_1 \models \Phi$$
$$E_0 \models \langle K\rangle\Phi \quad \text{iff} \quad \text{there is a run } E_0 \xrightarrow{a_1} E_1 \dots \text{ and } a_1 \in K \text{ and } E_1 \models \Phi$$

The operator $[-]$ expresses "next" over all runs and its dual $\langle - \rangle$ expresses weak "next."

A useful temporal operator is the binary until operator U. A finite or infinite length run $E_0 \xrightarrow{a_1} E_1 \xrightarrow{a_2} \dots$ satisfies the formula $\Phi \, \mathrm{U} \, \Psi$, "$\Phi$ is true until Ψ," provided there is an $i \geq 0$ such that $E_i \models \Psi$, and for each $j : 0 \leq j < i$ the intermediate process E_j has property Φ.

$$\begin{array}{ccccccc} E_0 & \xrightarrow{a_1} & E_1 & \xrightarrow{a_2} & \dots & E_i & \xrightarrow{a_{i+1}} \dots \\ \models & & \models & & \dots & \models & \\ \Phi & & \Phi & & & \Psi & \end{array}$$

The index i can be 0 (in which case Φ does not have to be true at any point of the run). If the run has zero length[3] then the index i must be zero. A special instance of U is when the first formula Φ is $\mathtt{tt}$. The formula $\mathtt{tt} \, \mathrm{U} \, \Psi$ expresses "eventually Ψ," which we abbreviate to $\mathrm{F}\Psi$.

A finite or infinite length run $E_0 \xrightarrow{a_1} E_1 \xrightarrow{a_2} \dots$ satisfies $\neg(\mathtt{tt} \, \mathrm{U} \, \neg\Psi)$[4], that is $\neg\mathrm{F}\neg\Psi$, if every process E_i within the run has the property Ψ.

$$\begin{array}{cccccccc} E_0 & \xrightarrow{a_1} & E_1 & \xrightarrow{a_2} & \dots & E_i & \xrightarrow{a_{i+1}} \dots \\ \models & & \models & & \dots & \models & \\ \Psi & & \Psi & & & \Psi & \dots \end{array}$$

This, therefore, expresses "always Ψ," which we abbreviate to $\mathrm{G}\Psi$. Notice that F and G are duals of each other.

Modal logic can be enriched by adding temporal operators to it. For each temporal operator (such as F) there are two variants, the strong variant ranging over all runs of a process, and the weak variant ranging over some run of the process. We preface the strong variant with A "for all runs" and the weak variant with E "for some run." Liveness and safety as described in the previous section

[3] The length of a run is the number of transitions within it.

[4] Here $\neg$ is the negation operator.

can now be properly defined.

$$\begin{array}{lcl} \text{Safety}(\Phi) & = & \text{AG}\neg\Phi \\ \text{Liveness}(\Phi) & = & \text{AF}\Phi \\ \text{WSafety}(\Phi) & = & \text{EG}\neg\Phi \\ \text{WLiveness}(\Phi) & = & \text{EF}\Phi \end{array}$$

If modal logic is extended with the two kinds of U operator the resulting temporal logic is a slight variant of computation tree temporal logic, CTL, due to Clarke, Emerson and Sistla [12]. We present the logic with an explicit negation operator.

$$\Phi ::= \texttt{tt} \mid \neg\Phi \mid \Phi_1 \wedge \Phi_2 \mid [K]\Phi \mid \text{A}(\Phi_1 \,\text{U}\, \Phi_2) \mid \text{E}(\Phi_1 \,\text{U}\, \Phi_2)$$

The definition of satisfaction between a process E_0 and a formula proceeds by induction on the formula. The only new clauses are for the two U operators that appeal to runs of E_0.

$$\begin{array}{lll} E_0 \models \text{A}(\Phi \,\text{U}\, \Psi) & \text{iff} & \text{for all runs } E_0 \xrightarrow{a_1} E_1 \xrightarrow{a_2} \ldots \text{ there is } i \geq 0 \\ & & \text{with } E_i \models \Psi \text{ and for all } j : 0 \leq j < i,\ E_j \models \Phi \\ E_0 \models \text{E}(\Phi \,\text{U}\, \Psi) & \text{iff} & \text{for some run } E_0 \xrightarrow{a_1} E_1 \xrightarrow{a_2} \ldots \text{ there is } i \geq 0 \\ & & \text{with } E_i \models \Psi \text{ and for all } j : 0 \leq j < i,\ E_j \models \Phi \end{array}$$

The two variants of "eventually" and "always" are definable as follows.

$$\begin{array}{lclclcl} \text{AF}\,\Phi & \stackrel{\text{def}}{=} & \text{A}(\texttt{tt}\,\text{U}\,\Phi) & \quad & \text{EF}\,\Phi & \stackrel{\text{def}}{=} & \text{E}(\texttt{tt}\,\text{U}\,\Phi) \\ \text{AG}\,\Phi & \stackrel{\text{def}}{=} & \neg\text{EF}\neg\Phi & & \text{EG}\,\Phi & \stackrel{\text{def}}{=} & \neg\text{AF}\neg\Phi \end{array}$$

Example 1 The level crossing has the crucial safety property that it is never possible for a train and a car to cross at the same time. In terms of runs, this means that no run of `Crossing` passes through a process that can perform both $\overline{\texttt{tcross}}$ and $\overline{\texttt{ccross}}$ as next actions, so the bad feature is $\langle\overline{\texttt{tcross}}\rangle\texttt{tt} \wedge \langle\overline{\texttt{ccross}}\rangle\texttt{tt}$. The safety property is therefore expressed by the CTL formula $\text{AG}([\overline{\texttt{tcross}}]\texttt{ff} \vee [\overline{\texttt{ccross}}]\texttt{ff})$.

Example 2 The weak liveness property of the slot machine $\texttt{SM}_n$ that it may eventually pay out a windfall is expressed as $\text{EF}\langle\overline{\texttt{win}}(10^6)\rangle\texttt{tt}$.

As with modal operators, temporal operators may be embedded within each other to express complex process features. An example is "eventually, action a is possible until b is always impossible," $\text{AF}(\langle a\rangle\texttt{tt}\,\text{U}\,\text{AG}[b]\texttt{ff})$.

In Section 4.1 it was shown that the ability to tick forever is not expressible in modal logic. It is also not directly expressible in CTL as defined here. For instance, the formula $\text{EG}\langle\texttt{tick}\rangle\texttt{tt}$ states that action `tick` is possible throughout some run, and process $\texttt{Cl}' \stackrel{\text{def}}{=} \texttt{tock.Cl}' + \texttt{tick.0}$ satisfies it. The problem is that the CTL temporal operators are not relativised to actions. A variant "always" operator is

G_K, which includes action information. A run satisfies $G_K \Phi$ if every transition in the run belongs to K and Φ is true throughout. The ability to tick forever is then directly expressible with the formula $EG_{\{\texttt{tick}\}}\langle - \rangle \texttt{tt}$, (where $\langle - \rangle \texttt{tt}$ ensures that the run is infinite).

In this work, we do not found temporal logic on runs. Partly this is because we wish to integrate action capabilities with temporal properties more elegantly than suggested above, partly because we wish to suppress the notion of a run. Instead we shall define appropriate closure conditions on sets of processes that express long term capabilities by appealing to inductive definitions built from modal logic. The idea is that a long term capability is just a particular closure of an immediate capability.

Exercises

1. Show the following
   - **a.** $\texttt{Ven} \models AF\langle \texttt{collect}_\texttt{b}, \texttt{collect}_\texttt{l} \rangle \texttt{tt}$
   - **b.** $\texttt{Ven} \models EF\langle \texttt{collect}_\texttt{b} \rangle \texttt{tt}$
   - **c.** $\texttt{Ven} \not\models AF\langle \texttt{collect}_\texttt{b} \rangle \texttt{tt}$
2. Show that $\texttt{Crossing} \models AG([\overline{\texttt{tcross}}]\texttt{ff} \vee [\overline{\texttt{ccross}}]\texttt{ff})$.
3. **a.** Show that CTL properties are preserved by bisimulation equivalence. That is, prove that if $E \sim F$, then for all $\Phi \in$ CTL, $E \models \Phi$ iff $F \models \Phi$.

   b. Let WCTL be CTL, except that the modal operators $[K]$ and $\langle K \rangle$ are replaced with the modalities of M^o of Section 2.4. Notice that the temporal operators of WCTL are still interpreted over runs involving thin transitions. Are WCTL properties preserved by observational equivalence $\approx$?
4. A run $E_0 \xrightarrow{a_1} E_1 \xrightarrow{a_2} \ldots$ satisfies $FG\Phi$ provided there is an $i \geq 0$ such that for all $j \geq i$, $E_j \models \Phi$.
   - **a.** Contrast the different meanings of the operators AFG and AFAG.
   - **b.** Give an example of a process E and a modal formula Φ such that $E \models AFG\Phi$ but $E \not\models AFAG\Phi$.
5. The clock $\texttt{Cl}_1 \stackrel{\text{def}}{=} \texttt{tick.tock.Cl}_1$ has the property that $\langle \texttt{tick} \rangle \texttt{tt}$ is true at every even point of its single run. Prove that the property "for every run $\langle \texttt{tick} \rangle \texttt{tt}$ is true at every even point" is not definable in CTL.

4.4 Modal formulas with variables

The modal logic M of Section 2.1 is extended with propositional variables which are ranged over by Z, as follows.

$$\Phi ::= Z \mid \texttt{tt} \mid \texttt{ff} \mid \Phi_1 \wedge \Phi_2 \mid \Phi_1 \vee \Phi_2 \mid [K]\Phi \mid \langle K \rangle \Phi$$

Modal formulas may contain propositional variables, and therefore the satisfaction relation $\models$ between a process and a formula needs to be refined. The important case is when a process has the property Z. Think of a propositional variable as a colour that can be ascribed to processes. A particular process may have a variety of different colours. Processes can be "coloured" arbitrarily. A particular colouring is defined by a "valuation" function V that assigns to each colour Z a subset of processes $\mathsf{V}(Z)$ having this colour. If there are two colours X, red, and Z, blue, then $\mathsf{V}(X)$ is the set of red coloured processes and $\mathsf{V}(Z)$ is the set of blue coloured processes.

The satisfaction relation between a process and a formula is relativised to a valuation. We write $E \models_{\mathsf{V}} \Phi$ when E has the property Φ relative to the valuation V, and $E \not\models_{\mathsf{V}} \Phi$ when E fails to have the property Φ relative to V. First is the semantic clause for a variable.

$$E \models_{\mathsf{V}} Z \text{ iff } E \in \mathsf{V}(Z)$$

Process E has colour Z relative to V iff E belongs to $\mathsf{V}(Z)$. The remaining semantic clauses are as in Section 2.1 for the logic M, except for the relativization to the colouring V. For example, the semantic clause for $\wedge$ is as follows.

$$E \models_{\mathsf{V}} \Phi_1 \wedge \Phi_2 \quad \text{iff} \quad E \models_{\mathsf{V}} \Phi_1 \text{ and } E \models_{\mathsf{V}} \Phi_2$$

The notation for the subset of E processes with property Φ, $\| \Phi \|^{\mathsf{E}}$, is also refined by an additional colouring component, $\| \Phi \|^{\mathsf{E}}_{\mathsf{V}}$.

$$\| \Phi \|^{\mathsf{E}}_{\mathsf{V}} \stackrel{\text{def}}{=} \{E \in \mathsf{E} : E \models_{\mathsf{V}} \Phi\}$$

The set E in $\| \Phi \|^{\mathsf{E}}_{\mathsf{V}}$ is invariably a transition closed set P or $\mathsf{P}(E)$, as defined in Section 1.6, and for each Z it is expected that the set of coloured processes $\mathsf{V}(Z)$ is a subset of E.

Valuations may be revised, so a colouring is then updated. A useful notation is $\mathsf{V}[\mathsf{E}/Z]$, which represents the valuation similar to V, except that E is the set of processes coloured Z.

$$(\mathsf{V}[\mathsf{E}/Z])(Y) = \begin{cases} \mathsf{E} & \text{if } Y = Z \\ \mathsf{V}(Y) & \text{otherwise} \end{cases}$$

There are many uses for revised valuations. Assume that Z does not occur in Φ. Whether $E \models_{\mathsf{V}} \Phi$ is independent of the colour Z. It follows that, for any colouring V and for any set E, $E \models_{\mathsf{V}} \Phi$ iff $E \models_{\mathsf{V}[\mathsf{E}/Z]} \Phi$. In particular, if Φ does not contain any variables (and is therefore also a formula of M), then whether E satisfies Φ is completely independent of colourings: in this special case we write $E \models \Phi$, as before.

A valuation V' extends the valuation V, which is written $\mathsf{V} \subseteq \mathsf{V}'$, if $\mathsf{V}(Z) \subseteq \mathsf{V}'(Z)$ for all variables Z. The colouring V' is uniformily more generous than V, so for any variable Z it follows that $\| Z \|_{\mathsf{V}}^{\mathsf{P}}$ is a subset of $\| Z \|_{\mathsf{V}'}^{\mathsf{P}}$. This feature extends to all formulas of M with variables.

Proposition 1 *If* V' *extends* V, *then* $\| \Phi \|_{\mathsf{V}}^{\mathsf{P}} \subseteq \| \Phi \|_{\mathsf{V}'}^{\mathsf{P}}$.

Proof. The proof proceeds by induction on Φ. We drop the index P. The base cases are when Φ is a variable Z, $\mathtt{tt}$ or $\mathtt{ff}$. The first of these cases follows directly from the definition of $\subseteq$ on valuations. Clearly, it also holds for $\mathtt{tt}$ and for $\mathtt{ff}$. The general case divides into the various subcases. First, suppose Φ is $\Psi_1 \wedge \Psi_2$, and assume that $\mathsf{V} \subseteq \mathsf{V}'$. By definition $\| \Phi \|_{\mathsf{V}}$ is the set $\{E : E \models_{\mathsf{V}} \Psi_1 \wedge \Psi_2\}$, which is just the set $\| \Psi_1 \|_{\mathsf{V}} \cap \| \Psi_2 \|_{\mathsf{V}}$. By the induction hypothesis $\| \Psi_i \|_{\mathsf{V}} \subseteq \| \Psi_i \|_{\mathsf{V}'}$ for $i = 1$ and $i = 2$. Therefore, $\| \Phi \|_{\mathsf{V}} \subseteq \| \Phi \|_{\mathsf{V}'}$. The case when Φ is $\Psi_1 \vee \Psi_2$ is similar. Next, assume that Φ is $[K]\Psi$. The set $\| \Phi \|_{\mathsf{V}}$ is therefore equal to $\{E : \text{if } E \stackrel{a}{\longrightarrow} F \text{ and } a \in K \text{ then } F \models_{\mathsf{V}} \Psi\}$, which is $\{E : \text{if } E \stackrel{a}{\longrightarrow} F \text{ and } a \in K \text{ then } F \in \| \Psi \|_{\mathsf{V}}\}$. By the induction hypothesis this is a subset of $\{E : \text{if } E \stackrel{a}{\longrightarrow} F \text{ and } a \in K \text{ then } F \in \| \Psi \|_{\mathsf{V}'}\}$, which is the definition of $\| \Phi \|_{\mathsf{V}'}$. The other modal case, Φ is $\langle K \rangle \Psi$, is similar and is left as an exercise. □

2^{P} denotes the set of all subsets of P. For instance, if P is $\{\mathtt{Cl}_1, \mathtt{tock.Cl}_1\}$, then 2^{P} is $\{\emptyset, \{\mathtt{Cl}_1\}, \{\mathtt{tock.Cl}_1\}, \mathsf{P}\}$. A function $g : 2^{\mathsf{P}} \to 2^{\mathsf{P}}$ maps elements of 2^{P}, subsets of P, into elements of 2^{P}. For any $\mathsf{E} \subseteq \mathsf{P}$, the element $g(\mathsf{E})$ is also a subset of P. With respect to a colour Z and valuation V, a modal formula Φ determines the function $f[\Phi, Z] : 2^{\mathsf{P}} \to 2^{\mathsf{P}}$, which when applied to the set $\mathsf{E} \subseteq \mathsf{P}$ is as follows.

$$f[\Phi, Z](\mathsf{E}) \stackrel{\text{def}}{=} \| \Phi \|_{\mathsf{V}[\mathsf{E}/Z]}^{\mathsf{P}} = \{E \in \mathsf{P} : E \models_{\mathsf{V}[\mathsf{E}/Z]} \Phi\}$$

Example 1 For any valuation V, the function $f[\langle \mathtt{tick} \rangle Z, Z]$ maps E into the set of processes that has a tick transition into E.

$$\begin{aligned} f[\langle \mathtt{tick} \rangle Z, Z](\mathsf{E}) &= \{E \in \mathsf{P} : E \models_{\mathsf{V}[\mathsf{E}/Z]} \langle \mathtt{tick} \rangle Z\} \\ &= \{E \in \mathsf{P} : \exists F.\, E \stackrel{\mathtt{tick}}{\longrightarrow} F \text{ and } F \models_{\mathsf{V}[\mathsf{E}/Z]} Z\} \\ &= \{E \in \mathsf{P} : \exists F \in \mathsf{E}.\, E \stackrel{\mathtt{tick}}{\longrightarrow} F\} \end{aligned}$$

Changing colour also changes the function. For any V, and $Y \neq Z$, $f[\langle \mathtt{tick} \rangle Z, Y]$ is a constant function.

$$\begin{aligned} f[\langle \mathtt{tick} \rangle Z, Y](\mathsf{E}) &= \{E \in \mathsf{P} : E \models_{\mathsf{V}[\mathsf{E}/Y]} \langle \mathtt{tick} \rangle Z\} \\ &= \{E \in \mathsf{P} : \exists F.\, E \stackrel{\mathtt{tick}}{\longrightarrow} F \text{ and } F \models_{\mathsf{V}[\mathsf{E}/Y]} Z\} \\ &= \{E \in \mathsf{P} : \exists F \in \mathsf{V}(Z).\, E \stackrel{\mathtt{tick}}{\longrightarrow} F\} \end{aligned}$$

A second example is that the function $f[[\mathtt{tick}]Z, Z]$ maps the set E to $\{E \in \mathsf{P} : \forall F.\, \text{if } E \stackrel{\mathtt{tick}}{\longrightarrow} F \text{ then } F \in \mathsf{E}\}$.

Example 2 Assume that Φ is a formula that does not contain the variable Z. For any V, the function $f[\Phi \vee \langle - \rangle Z, Z]$ maps E into the set of processes that have property Φ, or can do a transition into E.

$$\begin{aligned} f[\Phi \vee \langle - \rangle Z, Z](\mathsf{E}) &= \{E \in \mathsf{P} : E \models_{\mathsf{V}[\mathsf{E}/Z]} \Phi \vee \langle - \rangle Z\} \\ &= \{E \in \mathsf{P} : E \models_{\mathsf{V}[\mathsf{E}/Z]} \Phi \text{ or } E \models_{\mathsf{V}[\mathsf{E}/Z]} \langle - \rangle Z\} \\ &= \{E \in \mathsf{P} : E \models_{\mathsf{V}} \Phi \text{ or } \exists F \in \mathsf{E} \, \exists a. \, E \stackrel{a}{\longrightarrow} F\} \end{aligned}$$

Because Φ does not contain Z, $E \models_{\mathsf{V}[\mathsf{E}/Z]} \Phi$ iff $E \models_{\mathsf{V}} \Phi$.

With respect to P and V the set $f[\Phi, Z](\mathsf{E})$ can also be defined directly by induction on Φ. For example, the following covers the case when the formula is a conjunction.

$$f[\Phi_1 \wedge \Phi_2, Z](\mathsf{E}) = f[\Phi_1, Z](\mathsf{E}) \cap f[\Phi_2, Z](\mathsf{E})$$

A function $g : 2^{\mathsf{P}} \rightarrow 2^{\mathsf{P}}$ is "monotonic" with respect to the subset ordering, $\subseteq$, if it obeys the following condition.

$$\text{if } \mathsf{E} \subseteq \mathsf{F} \text{ then } g(\mathsf{E}) \subseteq g(\mathsf{F})$$

Proposition 1, above, has the consequence that, for any formula Φ and variable Z, the function $f[\Phi, Z]$ is monotonic with respect to any P and V.

Corollary 1 *For any* P *and* V, *the function* $f[\Phi, Z]$ *is monotonic.*

Proof. If $\mathsf{E} \subseteq \mathsf{F}$, then by definition $\mathsf{V}[\mathsf{E}/Z] \subseteq \mathsf{V}[\mathsf{F}/Z]$. It follows from Proposition 1 that $\| \Phi \|^{\mathsf{P}}_{\mathsf{V}[\mathsf{E}/Z]} \subseteq \| \Phi \|^{\mathsf{P}}_{\mathsf{V}[\mathsf{F}/Z]}$. However, $\| \Phi \|^{\mathsf{P}}_{\mathsf{V}[\mathsf{E}/Z]}$ is $f[\Phi, Z](\mathsf{E})$ and $\| \Phi \|^{\mathsf{P}}_{\mathsf{V}[\mathsf{F}/Z]}$ is $f[\Phi, Z](\mathsf{F})$, and therefore for any P and V, the function $f[\Phi, Z]$ is monotonic. □

Because V may be an arbitrary colouring, it is not in general true that, if $E \sim F$, then for any formula Φ, $E \models_{\mathsf{V}} \Phi$ iff $F \models_{\mathsf{V}} \Phi$. For instance, if $\mathsf{V}(Z)$ is the singleton set $\{E\}$ and $E \sim F$, then $E \models_{\mathsf{V}} Z$ but $F \not\models_{\mathsf{V}} Z$. However, if the colouring respects bisimulation equivalence, that is, if each set $\mathsf{V}(Z)$ of processes is bisimulation closed[5], then bisimilar processes will agree on properties. We extend the notion of bisimulation closure to colourings. The valuation V is bisimulation closed if, for each variable Z, the set $\mathsf{V}(Z)$ is bisimulation closed.

Proposition 2 *If* V *is bisimulation closed and* $E \sim F$, *then for all modal formulas* Φ *possibly containing variables,* $E \models_{\mathsf{V}} \Phi$ *iff* $F \models_{\mathsf{V}} \Phi$.

The proof of this result is a minor elaboration of the proof of Proposition 1 of Section 3.4, and is left as an exercise for the reader.

[5] E is bisimulation closed if $E \in \mathsf{E}$ and $F \in \mathsf{P}$, and $E \sim F$ implies $F \in \mathsf{E}$.

Exercises

1. Prove by induction on Φ that, for any set E if Z does not occur in Φ, then $E \models_{\mathsf{V}} \Phi$ iff $E \models_{\mathsf{V}[\mathsf{E}/Z]} \Phi$.
2. Show that, if for any variable Z, $\mathsf{V}'(Z) = \mathsf{V}(Z) \cap \mathsf{E}$, then the following is true: $\| \Phi \|_{\mathsf{V}}^{\mathsf{E}} = \| \Phi \|_{\mathsf{V}'}^{\mathsf{E}}$.
3. Relative to P and V, define the function $f[\Phi, Z]$ directly by induction on the structure of Φ.
4. If negation, $\neg$, is added to modal logic with variables, then show that Proposition 1 fails to hold, and that therefore there are functions $f[\Phi, Z]$ that are not monotonic.
5. Consider the extended modal logic with variables and the CTL temporal operators $\mathrm{A}(\Phi \mathrm{U} \Psi)$ and $\mathrm{E}(\Phi \mathrm{U} \Psi)$. Show that Proposition 1 still holds for this extended logic.
6. Assume that Φ and Ψ do not contain the variable Z. For any V and P, work out the following functions.
   - **a.** $f[\Phi \wedge [-]Z, Z]$
   - **b.** $f[\Phi \wedge \langle - \rangle Z, Z]$
   - **c.** $f[\Psi \vee (\Phi \wedge (\langle - \rangle \mathtt{tt} \wedge [-]Z)), Z]$
   - **d.** $f[\Psi \vee (\Phi \wedge \langle - \rangle Z), Z]$
   - **e.** $f[[a]((\langle b \rangle \mathtt{tt} \vee Y) \wedge Z), Z]$
   - **f.** $f[[a]((\langle b \rangle \mathtt{tt} \vee Y) \wedge Z), Y]$
7. Prove Proposition 2.

4.5 Modal equations and fixed points

Definitional equality, $\stackrel{\text{def}}{=}$, is essential for describing perpetual processes, as in the simplest case of the uncluttered clock `Cl`. Modal logic can be extended with this facility, following Larsen [36]. A modal equation has the form $Z \stackrel{\text{def}}{=} \Phi$, stipulating that Z expresses the same property as the formula Φ. The effect of this equation is to constrain the colour Z to processes having the property Φ. For instance, $Z \stackrel{\text{def}}{=} \langle \mathtt{tick} \rangle \mathtt{tt}$ stipulates that only processes that may immediately tick are coloured Z.

In the recursive modal equation $Z \stackrel{\text{def}}{=} \langle \mathtt{tick} \rangle Z$, both occurrences of Z select the same trait. What property is thereby expressed by Z? Recall from the previous section that $f[\langle \mathtt{tick} \rangle Z, Z]$ is a monotonic function that, when applied to argument $\mathsf{E} \subseteq \mathsf{P}$, is

$$\begin{aligned} f[\langle \mathtt{tick} \rangle Z, Z](\mathsf{E}) &= \| \langle \mathtt{tick} \rangle Z \|_{\mathsf{V}[\mathsf{E}/Z]}^{\mathsf{P}} \\ &= \{E \in \mathsf{P} : E \models_{\mathsf{V}[\mathsf{E}/Z]} \langle \mathtt{tick} \rangle Z\} \\ &= \{E \in \mathsf{P} : \exists F \in \mathsf{E}.\, E \stackrel{\mathtt{tick}}{\longrightarrow} F\}. \end{aligned}$$

Consequently, the recursive modal equation $Z \stackrel{\text{def}}{=} \langle \texttt{tick} \rangle Z$ constrains Z to be any set which obeys the following equality.

$$\begin{aligned} \mathsf{E} &= f[\langle \texttt{tick} \rangle Z, Z](\mathsf{E}) \\ &= \{E \in \mathsf{P} : \exists F \in \mathsf{E}.\, E \stackrel{\texttt{tick}}{\longrightarrow} F\} \end{aligned}$$

There may be many different subsets of P that are solutions. One example is the empty set

$$\emptyset = \{E \in \mathsf{P} : \exists F \in \emptyset.\, E \stackrel{\texttt{tick}}{\longrightarrow} F\}$$

because the right hand set must be empty. When P is the set $\{\texttt{Cl}\}$, then P is also a solution because of the transition $\texttt{Cl} \stackrel{\texttt{tick}}{\longrightarrow} \texttt{Cl}$.

$$\{\texttt{Cl}\} = \{E \in \{\texttt{Cl}\} : \exists F \in \{\texttt{Cl}\}.\, E \stackrel{\texttt{tick}}{\longrightarrow} F\}$$

A function $g : 2^{\mathsf{P}} \to 2^{\mathsf{P}}$ transforms subsets into subsets. $\mathsf{E} \subseteq \mathsf{P}$ is said to be a "fixed point" of g if the transformation leaves E unchanged, $g(\mathsf{E}) = \mathsf{E}$. Further applications of g also leave E fixed, $g(g(\mathsf{E})) = \mathsf{E}$, $g(g(g(\mathsf{E}))) = \mathsf{E}$, and so on. Every subset of P is a fixed point of the identity function. If g maps every subset into $\emptyset$, then it is the only fixed point. On the other hand, if g maps every set to a different set, so $g(\mathsf{E}) \neq \mathsf{E}$ for all E, then g does not have a fixed point. The fixed point constraint, $g(\mathsf{E}) = \mathsf{E}$, can be dissected.

E is a "prefixed point" of g, if $g(\mathsf{E}) \subseteq \mathsf{E}$

E is a "postfixed point" of g, if $\mathsf{E} \subseteq g(\mathsf{E})$

A fixed point has to be both a prefixed and a postfixed point. P is always a prefixed point, and $\emptyset$ is always a postfixed point.

A solution to the recursive modal equation $Z \stackrel{\text{def}}{=} \langle \texttt{tick} \rangle Z$ is therefore a fixed point of the function $f[\langle \texttt{tick} \rangle Z, Z]$. The definitions of prefixed and postfixed points can be viewed as "closure" conditions on putative solution sets E.

PRE $\quad f[\langle \texttt{tick} \rangle Z, Z](\mathsf{E}) \subseteq \mathsf{E}$

$\{E \in \mathsf{P} : \exists F \in \mathsf{E}.\, E \stackrel{\texttt{tick}}{\longrightarrow} F\} \subseteq \mathsf{E}$

if $E \in \mathsf{P}$ and $F \in \mathsf{E}$ and $E \stackrel{\texttt{tick}}{\longrightarrow} F$ then $E \in \mathsf{E}$

POST $\quad \mathsf{E} \subseteq f[\langle \texttt{tick} \rangle Z, Z](\mathsf{E})$

$\mathsf{E} \subseteq \{E \in \mathsf{P} : \exists F \in \mathsf{E}.\, E \stackrel{\texttt{tick}}{\longrightarrow} F\}$

if $E \in \mathsf{E}$ then $E \stackrel{\texttt{tick}}{\longrightarrow} F$ for some $F \in \mathsf{E}$

A fixed point must obey both closure conditions.

Example 1 If P is $\{\texttt{Cl}\}$, then both candidate sets $\emptyset$ and P obey the closure conditions PRE and POST, and are therefore fixed points of $f[\langle \texttt{tick} \rangle Z, Z]$. The solutions can be ordered by subset inclusion, $\emptyset \subseteq \{\texttt{Cl}\}$, offering a least and a greatest solution. In the case of the more sonorous clock $\texttt{Cl}_1$, which alternately ticks and tocks, there are more candidates for solutions, the sets $\emptyset$, $\{\texttt{Cl}_1\}$, $\{\texttt{tock.Cl}_1\}$, $\{\texttt{Cl}_1, \texttt{tock.Cl}_1\}$. Let f abbreviate $f[\langle\texttt{tick}\rangle Z, Z]$.

$$
\begin{array}{lcl}
f(\emptyset) & = & \emptyset \\
f(\{\texttt{Cl}_1\}) & = & \emptyset \\
f(\{\texttt{tock.Cl}_1\}) & = & \{\texttt{Cl}_1\} \\
f(\{\texttt{Cl}_1, \texttt{tock.Cl}_1\}) & = & \{\texttt{Cl}_1\}
\end{array}
$$

The prefixed points of f are $\emptyset$ and $\{\texttt{Cl}_1, \texttt{tock.Cl}_1\}$, and $\emptyset$ is the only postfixed point. Therefore, there is just a single fixed point in this example.

With respect to any set of processes P, the equation $Z \stackrel{\text{def}}{=} \langle\texttt{tick}\rangle Z$ has both a least and a greatest solution (which may coincide) with respect to the subset ordering. The general result guaranteeing these extremal solutions is due to Tarski and Knaster. It shows that the least solution is the intersection of all prefixed points, of all those subsets obeying PRE, and that the greatest solution is the union of all postfixed points, of all those subsets fulfilling POST. The result applies to arbitrary monotonic functions from subsets of P to subsets of P.

Proposition 1 *If $g : 2^{\mathsf{P}} \to 2^{\mathsf{P}}$ is a monotonic function with respect to $\subseteq$, then g*

1. *has a least fixed point given as the set $\bigcap\{\mathsf{E} \subseteq \mathsf{P} \,:\, g(\mathsf{E}) \subseteq \mathsf{E}\}$,*
2. *has a greatest fixed point given as the set $\bigcup\{\mathsf{E} \subseteq \mathsf{P} \,:\, \mathsf{E} \subseteq g(\mathsf{E})\}$.*

Proof. We show 1, leaving 2, which is proved by dual reasoning, as an exercise. Let E' be the set $\bigcap\{\mathsf{E} \subseteq \mathsf{P} \,:\, g(\mathsf{E}) \subseteq \mathsf{E}\}$. First, we establish that E' is indeed a fixed point, which means that $g(\mathsf{E}') = \mathsf{E}'$. Suppose $E \in g(\mathsf{E}')$. By definition $E \in g(\mathsf{E})$ for every $\mathsf{E} \subseteq \mathsf{P}$ such that $g(\mathsf{E}) \subseteq \mathsf{E}$. Consequently, $E \in \mathsf{E}$ for every such set E, so E belongs to their intersection too. This means that $g(\mathsf{E}') \subseteq \mathsf{E}'$. Next, assume that $E \in \mathsf{E}'$ but $E \notin g(\mathsf{E}')$. Let $\mathsf{E}_1 = \mathsf{E}' - \{E\}$. By monotonicity of g, $g(\mathsf{E}_1) \subseteq g(\mathsf{E}')$. But we have just shown $g(\mathsf{E}') \subseteq \mathsf{E}'$ and, since $E \notin g(\mathsf{E}')$, it follows that $g(\mathsf{E}') \subseteq \mathsf{E}_1$. Therefore, $g(\mathsf{E}_1) \subseteq \mathsf{E}_1$, which means that $\mathsf{E}' \subseteq \mathsf{E}_1$ by the definition of E', which is a contradiction. We have now shown that E' is a fixed point of g. Consider any other fixed point F. Because $g(\mathsf{F}) = \mathsf{F}$, it follows that $g(\mathsf{F}) \subseteq \mathsf{F}$, and by the definition of E' we know that $\mathsf{E}' \subseteq \mathsf{F}$, which means E' is the least fixed point. □

Proposition 1 guarantees that any recursive modal equation $Z \stackrel{\text{def}}{=} \Phi$ has extremal solutions, which are least and greatest fixed points of the monotonic function $f[\Phi, Z]$. Relinquishing the equational format, let $\mu Z.\,\Phi$ express the property given by the least fixed point of $f[\Phi, Z]$ and let $\nu Z.\,\Phi$ express the property determined by its greatest fixed point.

What properties are expressed by the extremal solutions of the modal equation $Z \stackrel{\text{def}}{=} \langle\texttt{tick}\rangle Z$? The least case, $\mu Z.\,\langle\texttt{tick}\rangle Z$, is of little import because it expresses the same property as $\texttt{ff}$[6]. More interesting is $\nu Z.\,\langle\texttt{tick}\rangle Z$, which expresses the longstanding ability to tick forever. This property is not expressible in modal logic, as was shown in Section 4.1. To see this, let $\mathsf{E} \subseteq \mathsf{P}$ consist of all those processes E_0 that have an infinite length run of the form, $E_0 \stackrel{\texttt{tick}}{\longrightarrow} E_1 \stackrel{\texttt{tick}}{\longrightarrow} \ldots$. Each process E_i mentioned in this run also belongs to E, so E obeys the closure condition POST earlier (that if $E \in \mathsf{E}$, then $E \stackrel{\texttt{tick}}{\longrightarrow} F$ for some $F \in \mathsf{E}$). The set determined by $\nu Z.\,\langle\texttt{tick}\rangle Z$ must therefore include E. Assume that there is a larger set $\mathsf{E}' \supset \mathsf{E}$ that also satisfies POST, and that F_0 belongs to the set $\mathsf{E}' - \mathsf{E}$. By the requirement POST, $F_0 \stackrel{\texttt{tick}}{\longrightarrow} F_1$ for some F_1 in E'. The process F_1 also belongs to $\mathsf{E}' - \mathsf{E}$ because, if it belonged to E, then F_0 would also be in E. The POST requirement can now be applied to F_1, so F_1 has a transition $F_1 \stackrel{\texttt{tick}}{\longrightarrow} F_2$ and F_2 in $\mathsf{E}' - \mathsf{E}$. Repeated application of this construction produces a perpetual run from F_0 of the form $F_0 \stackrel{\texttt{tick}}{\longrightarrow} F_1 \stackrel{\texttt{tick}}{\longrightarrow} \ldots$, where each $F_i \in \mathsf{E}' - \mathsf{E}$, but this is a contradiction, since each F_i can tick forever and therefore belongs to E. The ability to tick forever is therefore given as a simple closure condition of the immediate ability to tick.

Generalizing slightly the formula $\nu Z.\,\langle K\rangle Z$ expresses an ability to perform K actions forever. There are two special cases: $\nu Z.\,\langle -\rangle Z$ expresses a capacity for never ending behaviour, and $\nu Z.\,\langle \tau\rangle Z$ captures divergence, $\uparrow$ of Section 2.5, the ability to engage in infinite internal chatter.

A more composite recursive equation is $Z \stackrel{\text{def}}{=} \Phi \vee \langle -\rangle Z$. Assume that Φ does not contain Z. For P and V, the monotonic function $f[\Phi \vee \langle -\rangle Z, Z]$ applied to E is $\| \Phi \vee \langle -\rangle Z \|^{\mathsf{P}}_{\mathsf{V}[\mathsf{E}/Z]}$, which is

$$\{E \in \mathsf{P} : E \models_{\mathsf{V}} \Phi\} \cup \{E \in \mathsf{P} : \exists a \in \mathsf{A}.\, \exists F \in \mathsf{E}.\, E \stackrel{a}{\longrightarrow} F\}.$$

Because Φ does not contain Z, valuation V can be used instead of $\mathsf{V}[\mathsf{E}/Z]$ in the first subset. A fixed point E of this function is subject to the following closure conditions.

PRE if $E \in \mathsf{P}$ and $(E \models_{\mathsf{V}} \Phi$ or $\exists a \in \mathsf{A}.\, \exists F \in \mathsf{E}.\, E \stackrel{a}{\longrightarrow} F)$ then $E \in \mathsf{E}$

POST if $E \in \mathsf{E}$ then $E \models_{\mathsf{V}} \Phi$ or $\exists a \in \mathsf{A}.\, \exists F \in \mathsf{E}.\, E \stackrel{a}{\longrightarrow} F$

A subset E satisfying the condition PRE has to contain those processes with the property Φ. But then it also has to include processes F, that fail to satisfy Φ, but have a transition $F \stackrel{a}{\longrightarrow} E$, where $E \models_{\mathsf{V}} \Phi$. And so on. It turns out that a process E_0 has the property $\mu Z.\,\Phi \vee \langle -\rangle Z$ if there is a run $E_0 \stackrel{a_1}{\longrightarrow} E_1 \stackrel{a_2}{\longrightarrow} \ldots$ and an $i \geq 0$ and $E_i \models_{\mathsf{V}} \Phi$. That is, if E_0 has the weak eventually property $\mathrm{EF}\Phi$ of CTL. The largest solution also includes the extra possibility of performing actions forever without Φ ever becoming true, as the reader can check. A slight variant is to consider $\mu Z.\,\Phi \vee \langle K\rangle Z$, where K is a family of actions. This expresses that a process is able to perform K actions until Φ holds. When K is the singleton

[6] Because $\emptyset$ is always a fixed point of $f[\langle\texttt{tick}\rangle Z, Z]$.

set $\{\tau\}$, this formula expresses that, after some silent activity, Φ is true, expressed modally as $\langle\langle\ \rangle\rangle\Phi$.

Another example is the recursive equation $Z \stackrel{\text{def}}{=} \Psi \vee (\Phi \wedge \langle - \rangle Z)$, where neither Φ nor Ψ contains Z. The function $f[\Psi \vee (\Phi \wedge \langle - \rangle Z), Z]$ applied to E is the set

$$\{E : E \models_{\mathsf{V}} \Psi\} \cup \{E : E \models_{\mathsf{V}} \Phi \text{ and } \exists F \in \mathsf{E}.\exists a \in \mathsf{A}.\, E \stackrel{a}{\longrightarrow} F\}.$$

Its least fixed point with respect to P and V is the smallest set E that obeys the closure condition

$$(\{E : E \models_{\mathsf{V}} \Psi\} \cup \{E : E \models_{\mathsf{V}} \Phi \text{ and } \exists F \in \mathsf{E}.\exists a \in \mathsf{A}.\, E \stackrel{a}{\longrightarrow} F\}) \subseteq \mathsf{E}.$$

If $E \in \mathsf{P}$ and $E \models_{\mathsf{V}} \Psi$, then $E \in \mathsf{E}$. Also, if $E \models_{\mathsf{V}} \Phi$ and E has a transition $E \stackrel{a}{\longrightarrow} F$ with $F \in \mathsf{E}$, then $E \in \mathsf{E}$. Therefore, we can build up this least set E in stages as follows.

$$\begin{array}{lcl}
\mathsf{E}^1 & = & \{E : E \models_{\mathsf{V}} \Psi\} \\
\mathsf{E}^2 & = & \mathsf{E}^1 \cup \{E : E \models_{\mathsf{V}} \Phi \text{ and } \exists F \in \mathsf{E}^1.\exists a \in \mathsf{A}.\, E \stackrel{a}{\longrightarrow} F\} \\
\vdots & & \vdots \\
\mathsf{E}^{i+1} & = & \mathsf{E}^i \cup \{E : E \models_{\mathsf{V}} \Phi \text{ and } \exists F \in \mathsf{E}^i.\exists a \in \mathsf{A}.\, E \stackrel{a}{\longrightarrow} F\} \\
\vdots & & \vdots
\end{array}$$

The least set E will be the union of the sets E^i. Pictorially, the stages are as follows.

$$\begin{array}{cccccccc}
\mathbf{1} & & \mathbf{2} & & & \mathbf{i+1} & & \\
E & E & \stackrel{a_1}{\longrightarrow} & E_1 & E & \stackrel{a_1}{\longrightarrow} \ldots E_j \ldots \stackrel{a_i}{\longrightarrow} & & E_i \\
\models_{\mathsf{V}} & \models_{\mathsf{V}} & & \models_{\mathsf{V}} & \models_{\mathsf{V}} & \models_{\mathsf{V}} & & \models_{\mathsf{V}} \\
\Psi & \Phi & & \Psi & \Phi & \Phi & & \Psi
\end{array}$$

The formula $\mu Z.\, \Psi \vee (\Phi \wedge \langle - \rangle Z)$ captures the weak until of CTL, $\mathrm{E}(\Phi \mathrm{U} \Psi)$, as described in Section 4.3. Later, we shall describe the idea of computing stages more formally using approximants. The strong until $\mathrm{A}(\Phi \mathrm{U} \Psi)$ of CTL is defined as $\mu Z.\, \Psi \vee (\Phi \wedge (\langle - \rangle \mathtt{tt} \wedge [-]Z))$, where Z does not occur in Ψ or Φ. Later, we shall see that we can also define properties that are not expressible in CTL.

Exercises

1. Show that if $g : 2^{\mathsf{P}} \rightarrow 2^{\mathsf{P}}$ maps all arguments to different sets (that is, $g(\mathsf{E}) \neq \mathsf{E}$ for all E), then g is not monotonic.
2. Assume $h : 2^P \rightarrow 2^P$ is monotonic with respect to $\subseteq$. Prove the following.
   - **a.** if E_1 and E_2 are prefixed points of h, then $\mathsf{E}_1 \cap \mathsf{E}_2$ is also a prefixed point of h
   - **b.** if E_1 and E_2 are postfixed points of h, then $\mathsf{E}_1 \cup \mathsf{E}_2$ is also a postfixed point of h

3. Prove Proposition 1, part 2.
4. Assume $g : 2^{\mathsf{P}} \rightarrow 2^{\mathsf{P}}$ is an arbitrary function. The function g is inflationary if $\mathsf{E} \subseteq g(\mathsf{E})$ for any set $\mathsf{E} \subseteq \mathsf{P}$. Show that if g is inflationary, then g has fixed points.
5. What properties are expressed by the following formulas?
   a. $\nu Z.\, [\mathtt{tick}]Z$
   b. $\mu Z.\, [\mathtt{tick}]Z$
   c. $\nu Z.\, \langle\mathtt{tick}\rangle\langle-\rangle Z$
   d. $\nu Z.\, \langle\mathtt{tick}\rangle Z \wedge \langle\mathtt{tock}\rangle Z$
6. Show that $\nu Z.\, \Phi \vee \langle-\rangle Z$, when Φ does not contain Z, expresses the CTL property $\mathrm{EF}\Phi \vee \mathrm{EG}\langle-\rangle\mathtt{tt}$.
7. Contrast the properties expressed by the formulas $\mu Z.\, \Phi \wedge [\tau]Z$ and $\nu Z.\, \Phi \wedge [\tau]Z$ when Φ does not contain Z.
8. Assume that Φ and Ψ do not contain Z.
   a. What property does $\nu Z.\, \Psi \vee (\Phi \wedge \langle-\rangle Z)$ express?
   b. Show that $\mu Z.\, \Psi \vee (\Phi \wedge (\langle-\rangle\mathtt{tt} \wedge [-]Z))$ captures the strong until $\mathrm{A}(\Phi \mathrm{U} \Psi)$ of CTL.
   c. What does $\nu Z.\, \Psi \vee (\Phi \wedge (\langle-\rangle\mathtt{tt} \wedge [-]Z))$ express?
9. Prove that $\nu Z.\, \Phi \wedge [-]Z$ expresses $\mathrm{AG}\Phi$ (when Z does not occur in Φ).
10. What property does $\nu Z.\, \Phi \wedge [-][-]Z$ express, assuming that Φ does not contain Z? Prove that it is not expressible in CTL.
11. What property is expressed by the formula $\mu Y.\, \Phi \vee (\langle K\rangle Y \wedge \langle J\rangle Y)$ when Φ does not contain Y?

4.6 Duality

The expressive power of modal logic is increased when extended with least and greatest solutions of recursive modal equations. For instance, the least solution to $Z \stackrel{\mathrm{def}}{=} \Phi \vee \langle-\rangle Z$ (when Φ does not contain Z) expresses the weak liveness property $\mathrm{EF}\Phi$. The complement of weak liveness is safety. A process E does not satisfy $\mu Z.\, \Phi \vee \langle-\rangle Z$ if Φ never becomes true in any run of E.

Complements are directly expressible when negation is freely admitted into formulas as with CTL. The reason for avoiding negation in modal formulas is to preserve monotonicity. A simple example is the equation $Z \stackrel{\mathrm{def}}{=} \neg Z$. The function $f[\neg Z, Z]$ is not monotonic because $f[\neg Z, Z](\mathsf{E}) = \mathsf{P} - \mathsf{E}$. A fixed point E of this function must obey $\mathsf{E} = \mathsf{P} - \mathsf{E}$, which is impossible because P is non-empty. Another example is $Z \stackrel{\mathrm{def}}{=} \langle\mathtt{tock}\rangle\mathtt{tt} \wedge [\mathtt{tick}]\neg Z$. The function $f[\langle\mathtt{tock}\rangle\mathtt{tt} \wedge$

$[\texttt{tick}]\neg Z, Z]$ applied to E is the set

$$\{E \in \mathsf{P} : \exists F.\, E \stackrel{\texttt{tock}}{\longrightarrow} F\} \cap \{E \in \mathsf{P} : \forall F. \text{if } E \stackrel{\texttt{tick}}{\longrightarrow} F \text{ then } F \notin \mathsf{E}\}.$$

In general, this function is not monotonic, and whether it has fixed points depends on the structure of the set P.

Negation can be admitted into modal formulas provided the following restriction is placed on the form of a recursive modal equation $Z \stackrel{\text{def}}{=} \Phi$: every free occurrence of Z in Φ lies within the scope of an even number of negations. This guarantees the function $f[\Phi, Z]$ is monotonic. The two examples above do not comply with this condition.

The complement of a formula is also in the logic without the explicit presence of negation. This was shown for modal formulas of M in Section 2.2 where Φ^c, the complement of Φ, is defined inductively as follows.

$$\begin{array}{lcllcl} \texttt{tt}^c & = & \texttt{ff} & \texttt{ff}^c & = & \texttt{tt} \\ (\Phi \wedge \Psi)^c & = & \Phi^c \vee \Psi^c & (\Phi \vee \Psi)^c & = & \Phi^c \wedge \Psi^c \\ ([K]\Phi)^c & = & \langle K\rangle\Phi^c & (\langle K\rangle\Phi)^c & = & [K]\Phi^c \end{array}$$

It also turns out that the fixed point operators are duals of each other.

$$\begin{array}{lcl} Z^c & = & Z \\ (\nu Z.\,\Phi)^c & = & \mu Z.\,\Phi^c \\ (\mu Z.\,\Phi)^c & = & \nu Z.\,\Phi^c \end{array}$$

Assume that $\mathsf{V}(Z) \subseteq \mathsf{P}$ for all Z. The complement valuation V^c with respect to P is given as $\mathsf{V}^c(Z) = \mathsf{P} - \mathsf{V}(Z)$ for all Z. The following result shows that Φ^c is the complement of Φ modulo complementation of V.

Proposition 1 $E \models_{\mathsf{V}} \Phi$ *iff* $E \not\models_{\mathsf{V}^c} \Phi^c$.

Proof. This result generalises Proposition 1 of Section 2.2 and is proved by structural induction on Φ. The base cases are when Φ is $\texttt{tt}$, $\texttt{ff}$, or Z. The first two are clear. $E \models_{\mathsf{V}} Z$ iff $E \in \mathsf{V}(Z)$ iff $E \notin \mathsf{V}^c(Z)$ iff $E \not\models_{\mathsf{V}^c} Z$. For the inductive step, the boolean and modal cases are as in Proposition 1 of Section 2.2. The new cases are those involving fixed points.

$$\begin{array}{lll} E \models_{\mathsf{V}} \nu Z.\,\Phi & \text{iff} & E \in \bigcup\{\mathsf{E} \subseteq \mathsf{P} : \mathsf{E} \subseteq \|\,\Phi\,\|^{\mathsf{P}}_{\mathsf{V}[\mathsf{E}/Z]}\} \\ & \text{iff} & E \notin \mathsf{P} - \bigcup\{\mathsf{E} \subseteq \mathsf{P} : \mathsf{E} \subseteq \|\,\Phi\,\|^{\mathsf{P}}_{\mathsf{V}[\mathsf{E}/Z]}\} \\ & \text{iff} & E \notin \bigcap\{\mathsf{P} - \mathsf{E} : \mathsf{E} \subseteq \|\,\Phi\,\|^{\mathsf{P}}_{\mathsf{V}[\mathsf{E}/Z]}\}, \end{array}$$

which by the induction hypothesis is as follows.

$$\begin{array}{ll} \text{iff} & E \notin \bigcap\{\mathsf{P} - \mathsf{E} : \|\,\Phi^c\,\|^{\mathsf{P}}_{\mathsf{V}^c[\mathsf{P}-\mathsf{E}/Z]} \subseteq \mathsf{P} - \mathsf{E}\} \\ \text{iff} & E \notin \bigcap\{\mathsf{E} \subseteq \mathsf{P} : \|\,\Phi^c\,\|^{\mathsf{P}}_{\mathsf{V}^c[\mathsf{E}/Z]} \subseteq \mathsf{E}\} \\ \text{iff} & E \not\models_{\mathsf{V}^c} \mu Z.\,\Phi^c \\ \text{iff} & E \not\models_{\mathsf{V}^c} (\nu Z.\,\Phi)^c \end{array}$$

The other case $E \models_V \mu Z.\,\Phi$ is similar and is left as an exercise. □

If a property is an extremal solution to the equation $Z \stackrel{\text{def}}{=} \Phi$, then its complement is the dual solution to the equation $Z \stackrel{\text{def}}{=} \Phi^c$. Consider convergence, $E\downarrow$, which holds if E is unable to perform silent actions for ever. The formula $\nu Z.\,\langle\tau\rangle Z$ expresses its complement, divergence. Hence, convergence is defined as the *least* solution to the equation $Z \stackrel{\text{def}}{=} (\langle\tau\rangle Z)^c$, which is $\mu Z.\,[\tau]Z$.

Safety is the complement of weak liveness. The formula $\mu Z.\,\Phi \vee \langle -\rangle Z$ captures $\mathrm{EF}\Phi$. Its complement is the largest solution to $Z \stackrel{\text{def}}{=} (\Phi \vee \langle -\rangle Z)^c$, which is $\nu Z.\,\Phi^c \wedge [-]Z$. This formula expresses "Φ is never true," $\mathrm{AG}\Phi^c$.

Example 1 The level crossing of Figure 1.10 has the crucial safety property that it is never possible for a train and a car to cross at the same time. The feature to be avoided is $\langle\overline{\mathtt{tcross}}\rangle\mathtt{tt} \wedge \langle\overline{\mathtt{ccross}}\rangle\mathtt{tt}$, so the safety property is given by $\nu Z.\,([\overline{\mathtt{tcross}}]\mathtt{ff} \vee [\overline{\mathtt{ccross}}]\mathtt{ff}) \wedge [-]Z$.

The slot machine eventually produces winnings or an indication of loss. A slightly better description is "whenever a coin is input, eventually either a loss or a winning sum of money is output." This property is expressed using both fixed point operators. Embedding fixed point operators within each other goes beyond the simple equational format here described.

Exercises

1. Let f be the function $f[\langle\mathtt{tock}\rangle\mathtt{tt} \wedge [\mathtt{tick}]\neg Z, Z]$.
   a. Give a set P such that f does not have fixed points
   b. Give a set P such that f has both least and greatest fixed points
2. Assume that formulas may contain occurrences of $\neg$. Prove that, if every occurrence of Z within Φ lies within the scope of an even number of negations, then the recursive equation $Z \stackrel{\text{def}}{=} \Phi$ has both a least and a greatest solution.
3. When $\neg$ is freely admitted into formulas, show the following (where two formulas Φ and Ψ are equivalent if for all processes E, $E \models \Phi$ iff $E \models \Psi$).
   a. $\mu Z.\,\Phi$ is equivalent to $\neg\nu Z.\,\neg\Phi\{\neg Z/Z\}$
   b. $\nu Z.\,\Phi$ is equivalent to $\neg\mu Z.\,\neg\Phi\{\neg Z/Z\}$
4. What property of processes does $\mu Z.\,[-]Z$ express?
5. Prove that $[\![\ ]\!]\,\Phi$ is expressed as the formula $\nu Z.\,\Phi \wedge [\tau]Z$.

5

Modal Mu-Calculus

In the previous chapter we saw that modal formulas are not very expressive. They can not capture enduring traits of processes, the properties definable within temporal logic. However, these longer term properties can be viewed as closure conditions on immediate capabilities and necessities that modal logic captures. By permitting recursive modal equations, these temporal properties are expressible as extremal solutions of such equations. The property "whenever a coin is inserted, eventually an item is collected" is expressed using two recursive modal equations with different solutions. In the previous chapter, least and greatest solutions to recursive modal equations were represented using the fixed point quantifiers μZ and νZ. In this chapter we shall explicitly add these connectives to modal logic, thereby providing a very rich temporal logic.

5.1 Modal logic with fixed points

Modal logic with the extremal fixed point operators νZ and μZ is known as "modal mu-calculus," μM. Formulas of μM are built from variables, boolean connectives, modal operators and the fixed point operators.

$$\Phi \;::=\; \mathtt{tt} \mid \mathtt{ff} \mid Z \mid \Phi_1 \wedge \Phi_2 \mid \Phi_1 \vee \Phi_2 \mid [K]\Phi \mid \langle K\rangle\Phi \mid \nu Z.\,\Phi \mid \mu Z.\,\Phi$$

In the sequel, we let σ range over the set $\{\mu, \nu\}$. Formulas of μM may contain multiple occurrences of fixed point operators. An occurrence of a variable Z is "free" within a formula if it is not within the scope of an occurrence of σZ. The operator σZ in the formula $\sigma Z.\Phi$ is a quantifier that binds free occurrences of Z in Φ. We assume that σZ has wider scope than other operators. The scope of μZ is the rest of the formula in

$$\mu Z.\,\langle J\rangle Y \vee (\langle b\rangle Z \wedge \mu Y.\,\nu Z.\,([b]Y \wedge [K]Z))$$

and it binds the occurrence of Z in the subformula $\langle b\rangle Z$, but does not bind the occurrence of Z in $[K]Z$, which is bound by the occurrence of νZ. There is just one free variable occurrence in this formula, that of Y in $\langle J\rangle Y$. An occurrence of σZ may bind more than one occurrence of Z, as in $\nu Z.\,\langle \mathtt{tick}\rangle Z \wedge \langle \mathtt{tock}\rangle Z$.

The satisfaction relation $\models_{\mathsf{V}}$ between processes and formulas (relative to the valuation V) is defined inductively on the structure of formulas. First, are the cases that have been presented previously.

$$\begin{array}{lll}
E \models_{\mathsf{V}} \mathtt{tt} & & \\
E \not\models_{\mathsf{V}} \mathtt{ff} & & \\
E \models_{\mathsf{V}} Z & \text{iff} & E \in \mathsf{V}(Z) \\
E \models_{\mathsf{V}} \Phi \wedge \Psi & \text{iff} & E \models_{\mathsf{V}} \Phi \text{ and } E \models_{\mathsf{V}} \Psi \\
E \models_{\mathsf{V}} \Phi \vee \Psi & \text{iff} & E \models_{\mathsf{V}} \Phi \text{ or } E \models_{\mathsf{V}} \Psi \\
E \models_{\mathsf{V}} [K]\Phi & \text{iff} & \forall F \in \{E' : E \xrightarrow{a} E' \text{ and } a \in K\}.\ F \models_{\mathsf{V}} \Phi \\
E \models_{\mathsf{V}} \langle K\rangle\Phi & \text{iff} & \exists F \in \{E' : E \xrightarrow{a} E' \text{ and } a \in K\}.\ F \models_{\mathsf{V}} \Phi
\end{array}$$

The remaining cases are the fixed point operators, and we appeal to sets of processes, $\| \Phi \|^{\mathsf{P}}_{\mathsf{V}}$, as defined in Section 4.4, $\{E \in \mathsf{P} : E \models_{\mathsf{V}} \Phi\}$, in their definition. It is assumed in both cases that E belongs to P.

$$\begin{array}{lll}
E \models_{\mathsf{V}} \nu Z.\,\Phi & \text{iff} & E \in \bigcup\{\mathsf{E} \subseteq \mathsf{P} : \mathsf{E} \subseteq \| \Phi \|^{\mathsf{P}}_{\mathsf{V}[\mathsf{E}/Z]}\} \\
E \models_{\mathsf{V}} \mu Z.\,\Phi & \text{iff} & E \in \bigcap\{\mathsf{E} \subseteq \mathsf{P} : \| \Phi \|^{\mathsf{P}}_{\mathsf{V}[\mathsf{E}/Z]} \subseteq \mathsf{E}\}
\end{array}$$

These clauses are instances of Proposition 1 of Section 4.5. A greatest fixed point is the union of postfixed points, and a least fixed point is the intersection of prefixed points. To justify their use here we need to show that, for any variable Z and μM formula Φ, the function $f[\Phi, Z]$ is monotonic[1]. The formula Φ may contain fixed points, and therefore we need to extend Proposition 1 of Section 4.4 from modal formulas to μM formulas. Recall that the valuation V' extends V, written $\mathsf{V} \subseteq \mathsf{V}'$, if for each variable Z, $V(Z) \subseteq V'(Z)$.

Proposition 1 *If* V' *extends* V, *then* $\| \Phi \|_{\mathsf{V}}^{\mathsf{P}} \subseteq \| \Phi \|_{\mathsf{V}'}^{\mathsf{P}}$.

Proof. We show by induction on Φ that, if $\mathsf{V} \subseteq \mathsf{V}'$, then $\| \Phi \|_{\mathsf{V}}^{\mathsf{P}} \subseteq \| \Phi \|_{\mathsf{V}'}^{\mathsf{P}}$. We now drop the index P. The base cases, boolean cases and modal cases are exactly as expressed in the proof of Proposition 1 of Section 4.4. This just leaves the fixed point cases. Let Φ be the formula $\nu Y. \Psi$. Suppose $E \in \| \Phi \|_{\mathsf{V}}$. By the semantic clause for νY, there is a set E containing E with the property that $\mathsf{E} \subseteq \| \Psi \|_{\mathsf{V}[\mathsf{E}/Y]}$. Because $\mathsf{V} \subseteq \mathsf{V}'$, it follows that $\mathsf{V}[\mathsf{E}/Y] \subseteq \mathsf{V}'[\mathsf{E}/Y]$ and therefore by the induction hypothesis $\| \Psi \|_{\mathsf{V}[\mathsf{E}/Y]} \subseteq \| \Psi \|_{\mathsf{V}'[\mathsf{E}/Y]}$. Consequently, $\mathsf{E} \subseteq \| \Psi \|_{\mathsf{V}'[\mathsf{E}/Y]}$, and therefore $E \in \| \Phi \|_{\mathsf{V}'}$ too. The other case when Φ is $\mu Y. \Psi$ is similar and is left as an exercise. □

A straightforward corollary of Proposition 1 is that, for any μM formula Φ and valuation V if $\mathsf{E} \subseteq \mathsf{F}$, then $\| \Phi \|_{\mathsf{V}[\mathsf{E}/Z]}^{\mathsf{P}} \subseteq \| \Phi \|_{\mathsf{V}[\mathsf{F}/Z]}^{\mathsf{P}}$, and that therefore the function $f[\Phi, Z]$ is monotonic.

A slightly different presentation of the clauses for the fixed points dispenses with explicit use of sets $\| \Phi \|_{\mathsf{V}}^{\mathsf{P}}$.

$E \models_{\mathsf{V}} \nu Z. \Phi$	iff	$\exists \mathsf{E} \subseteq \mathsf{P}.\ E \in \mathsf{E}$ and $\forall F \in \mathsf{E}.\ F \models_{\mathsf{V}[\mathsf{E}/Z]} \Phi$
$E \models_{\mathsf{V}} \mu Z. \Phi$	iff	$\forall \mathsf{E} \subseteq \mathsf{P}.$ if $E \notin \mathsf{E}$ then $\exists F \in \mathsf{P}.\ F \models_{\mathsf{V}[\mathsf{E}/Z]} \Phi$ and $F \notin \mathsf{E}$

The first is a simple reformulation of the clause above, and the second follows by routine calculation:

$$
\begin{array}{lll}
E \models_{\mathsf{V}} \mu Z. \Phi & \text{iff} & E \in \bigcap \{\mathsf{E} \subseteq \mathsf{P} : \| \Phi \|_{\mathsf{V}[\mathsf{E}/Z]} \subseteq \mathsf{E}\} \\
& \text{iff} & \forall \mathsf{E} \subseteq \mathsf{P}. \text{ if } \| \Phi \|_{\mathsf{V}[\mathsf{E}/Z]} \subseteq \mathsf{E} \text{ then } E \in \mathsf{E} \\
& \text{iff} & \forall \mathsf{E} \subseteq \mathsf{P}. \text{ if } E \notin \mathsf{E} \text{ then } \| \Phi \|_{\mathsf{V}[\mathsf{E}/Z]} \not\subseteq \mathsf{E} \\
& \text{iff} & \forall \mathsf{E} \subseteq \mathsf{P}. \text{ if } E \notin \mathsf{E} \text{ then } \exists F \in \mathsf{P}.\ F \models_{\mathsf{V}[\mathsf{E}/Z]} \Phi \text{ and } F \notin \mathsf{E}.
\end{array}
$$

An "unfolding" of a fixed point formula $\sigma Z. \Phi$ is the formula $\Phi\{\sigma Z.\Phi/Z\}$: the fixed point formula is substituted for all free occurrences of Z in the "body" Φ. For instance, the unfolding of $\nu Z. \langle - \rangle Z$ is $\langle - \rangle (\nu Z. \langle - \rangle Z)$. The meaning of a fixed point formula is the same as its unfolding.

Proposition 2 $E \models_{\mathsf{V}} \sigma Z.\Phi$ *iff* $E \models_{\mathsf{V}} \Phi\{\sigma Z.\Phi/Z\}$.

[1]Relative to P and V, for any $\mathsf{E} \subseteq \mathsf{P}$, $f[\Phi, Z](\mathsf{E})$ is the set $\| \Phi \|_{\mathsf{V}[\mathsf{E}/Z]}^{\mathsf{P}}$.

Modal mu-calculus was originally proposed by Kozen [34] (and see also Pratt [50]) as an extension of propositional dynamic logic[2]. Its roots lie in more general program logics with extremal fixed points, originally developed by Park, De Bakker and De Roever. Larsen suggested that Hennessy-Milner logic with fixed points is useful for describing properties of processes [36]. Previously, Clarke and Emerson used extremal fixed points on top of a temporal logic for expressing properties of concurrent systems [18].

Example 1 The vending machine Ven of Section 1.1 has the property "whenever a coin is inserted, eventually an item is collected," expressed as $\nu Z.\, [\mathtt{2p}, \mathtt{1p}]\Psi \wedge [-]Z$, when Ψ is $\mu Y.\, \langle - \rangle \mathtt{tt} \wedge [-\{\mathtt{collect}_b, \mathtt{collect}_l\}]Y$. The appropriate set P is

$$\{\mathtt{Ven}, \mathtt{Ven_b}, \mathtt{Ven_l}, \mathtt{collect_b.Ven}, \mathtt{collect_l.Ven}\}.$$

Let V be any valuation. First we show that $\| \Psi \|^{\mathsf{P}}_{\mathsf{V}}$ is the full set P. Clearly, P is a prefixed point.

$$\| \langle - \rangle \mathtt{tt} \wedge [-\{\mathtt{collect_b}, \mathtt{collect_l}\}]Y \|^{\mathsf{P}}_{\mathsf{V}[\mathsf{P}/Z]} \subseteq \mathsf{P}$$

It is in fact the smallest prefixed point. Any proper subset E of P fails the associated closure condition, where A′ is the set $\mathsf{A} - \{\mathtt{collect_b}, \mathtt{collect_l}\}$.

$$\text{If } (\exists F.\exists a.\, E \stackrel{a}{\longrightarrow} F \text{ and } \forall F.\forall a \in \mathsf{A}'.E \stackrel{a}{\longrightarrow} F \text{ implies } F \in \mathsf{E}), \text{ then } E \in \mathsf{E}$$

For example, if E is the subset $\mathsf{P} - \{\mathtt{Ven_b}\}$, then $\mathtt{Ven_b}$ satisfies the antecedent of this closure condition, and therefore E is not a prefixed point. Similarly, $\emptyset$ fails to be a prefixed point because $\mathtt{collect_b.Ven}$ satisfies the antecedent. Given that $\| \Psi \|^{\mathsf{P}}_{\mathsf{V}}$ is P, it follows that $\| \nu Z.\, [\mathtt{2p}, \mathtt{1p}]\Psi \wedge [-]Z \|^{\mathsf{P}}_{\mathsf{V}}$ is also P.

As with modal formulas, a formula Φ is said to be realizable (or satisfiable) if there is a process that satisfies it. For example, the clock Cl realizes $\nu Z.\, \langle \mathtt{tick} \rangle Z$. In contrast $\mu Z.\, \langle \mathtt{tick} \rangle Z$ is not satisfiable. There is a technique for deciding whether a formula is realizable, due to Streett and Emerson [56]. An important consequence of their proof is that modal mu-calculus has the "finite model property:" if a formula holds of a process then there is a finite state process satisfying it.

Proposition 3 *If $E \models_{\mathsf{V}} \Phi$, then there is a finite state process F and valuation V′ such that $F \models_{\mathsf{V}'} \Phi$.*

Exercises

1. Show the following
   a. $\mathtt{Cl} \models \nu Z.\, \langle \mathtt{tick} \rangle Z \vee [\mathtt{tick}]\mathtt{ff}$
   b. $\mathtt{tick.0} \models \nu Z.\, \langle \mathtt{tick} \rangle Z \vee [\mathtt{tick}]\mathtt{ff}$

---

[2] The modalities of μM slightly extend those of Kozen's logic because sets of labels may appear within them instead of single labels, and on the other hand Kozen has explicit negation. Kozen calls the logic "propositional mu-calculus," which would be more appropriate to boolean logic with fixed points.

c. $\texttt{Cl} \not\models \mu Z.\, \langle \texttt{tick} \rangle Z \vee [\texttt{tick}]\texttt{ff}$

d. $\texttt{tick.0} \models \mu Z.\, \langle \texttt{tick} \rangle Z \vee [\texttt{tick}]\texttt{ff}$

2. Let $\mathsf{P} = \mathsf{P}(\texttt{Ven})$. Determine the sets $\| \Phi \|_{\mathsf{V}}^{\mathsf{P}}$ when Φ is each of the following formulas.

   a. $\mu Z.\, \langle \texttt{2p}, \texttt{1p} \rangle \texttt{tt} \vee [-]Z$

   b. $\mu Z.\, \langle \texttt{little} \rangle \texttt{tt} \vee [-]Z$

   c. $\mu Z.\, \langle \texttt{little} \rangle \texttt{tt} \vee [-]Z$

   d. $\nu Z.\, [\texttt{2p}](\mu Y.\, \langle \texttt{collect}_{\texttt{b}} \rangle \texttt{tt} \vee [-]Y) \wedge [-]Z$

3. Assume that $\texttt{D}$ and $\texttt{D}'$ are the following two processes $\texttt{D} \stackrel{\text{def}}{=} a.\texttt{D}'$ and $\texttt{D}' \stackrel{\text{def}}{=} b.0 + a.\texttt{D}$. Show the following.

   a. $D \models \nu Z.\, \mu Y.\, [a]((\langle b \rangle \texttt{tt} \wedge Z) \vee Y)$

   b. $D' \models \nu Z.\, \mu Y.\, [a]((\langle b \rangle \texttt{tt} \wedge Z) \vee Y)$

   c. $D \not\models \mu Y.\, \nu Z.\, [a]((\langle b \rangle \texttt{tt} \vee Y) \wedge Z)$

   d. $0 \models \mu Y.\, \nu Z.\, [a]((\langle b \rangle \texttt{tt} \vee Y) \wedge Z)$

4. Show that, if Z is not free in Φ, then $E \models_{\mathsf{V}} \Phi$ iff $E \models_{\mathsf{V}[\emptyset/Z]} \Phi$.

5. a. Carefully define the substitution operation $\Phi\{\Psi/Z\}$ by induction on Φ.

   b. Prove Proposition 2, above.

6. In propositional dynamic logic there is some structure on actions.

$$E \stackrel{w;v}{\longrightarrow} F \quad \text{iff} \quad E \stackrel{w}{\longrightarrow} E_1 \stackrel{v}{\longrightarrow} F \text{ for some } E_1$$

$$E \stackrel{w^*}{\longrightarrow} F \quad \text{iff} \quad E = F \text{ or } E \stackrel{w}{\longrightarrow} E_1 \stackrel{w}{\longrightarrow} \dots \stackrel{w}{\longrightarrow} E_n \stackrel{w}{\longrightarrow} F \text{ for some } n \geq 0 \text{ and } E_1, \dots, E_n$$

   Show the following, assuming that Φ does not contain Z as a free variable.

   a. $E \models_{\mathsf{V}} [a;b]\Phi$ iff $E \models_{\mathsf{V}} [a][b]\Phi$

   b. $E \models_{\mathsf{V}} [a^*]\Phi$ iff $E \models_{\mathsf{V}} \nu Z.\, \Phi \wedge [a]Z$

7. What properties are expressed by the following formulas?

   a. $\mu Y.\, [b]\texttt{ff} \wedge [a](\mu Z.\, [a]\texttt{ff} \wedge [b]Y \wedge [-b]Z) \wedge [-a]Y$

   b. $\nu Y.\, [b]\texttt{ff} \wedge [a](\mu Z.\, [a]\texttt{ff} \wedge [b]Y \wedge [-b]Z) \wedge [-a]Y$

   c. $\mu X.\, \nu Y.\, [a]X \wedge [-a]Y$

   d. $\nu Z.\, [a](\mu Y.\, \langle - \rangle \texttt{tt} \wedge [-b]Y) \wedge [-]Z$

   e. $\nu Z.\, (\mu X.\, [b](\nu Y.\, [c](\nu Y_1.\, X \wedge [-a]Y_1) \wedge [-a]Y) \wedge [-]Z)$

5.2 Macros and normal formulas

A common complaint about μM is that formulas can be difficult to understand, and that it can be hard to find the right formula to express a particular property.

Of course, this is also true of almost any notation involving binding or embedded operators. For example, it is not immediately clear what property the following CTL formula $\mathrm{AG}(\mathrm{EF}\langle \mathtt{tick}\rangle \mathtt{tt} \wedge \mathrm{AF}[\mathtt{tock}]\mathtt{ff})$ expresses. However, the problem is more acute in the case of μM because of fixed points.

Because μM is very expressive, we can introduce a variety of macros. For example, as we saw in Chapter 4, the temporal operators of CTL are definable as fixed points in a straightforward fashion (where Φ and Ψ do not contain Z).

$$\begin{aligned} \mathrm{A}(\Phi \mathrm{\,U\,} \Psi) &\equiv \mu Z.\, \Psi \vee (\Phi \wedge (\langle - \rangle \mathtt{tt} \wedge [-]Z)) \\ \mathrm{E}(\Phi \mathrm{\,U\,} \Psi) &\equiv \mu Z.\, \Psi \vee (\Phi \wedge \langle - \rangle Z) \end{aligned}$$

Macros can also be introduced for the starring operator of propositional dynamic logic (where again Φ does not contain Z).

$$[K^*]\Phi \equiv \nu Z.\, \Phi \wedge [K]Z$$

Formulas of μM exhibit a duality. A formula Φ that does not contain occurrences of free variables has a straightforward complement Φ^c, as defined in Section 4.6. The following is an example.

$$(\nu Z.\, \mu Y.\, \nu X.\, [a]((\langle b \rangle X \wedge Z) \vee [K]Y))^c = \mu Z.\, \nu Y.\, \mu X.\, \langle a \rangle(([b]X \vee Z) \wedge \langle K \rangle Y)$$

Because CTL has explicit negation, we can also introduce macros for negations of the two kinds of until formulas as follows (where again Z does not occur in Φ or Ψ).

$$\begin{aligned} \neg\mathrm{A}(\Phi \mathrm{\,U\,} \Psi) &\equiv \nu Z.\, \Psi^c \wedge (\Phi^c \vee ([-]\mathtt{ff} \vee \langle - \rangle Z)) \\ \neg\mathrm{E}(\Phi \mathrm{\,U\,} \Psi) &\equiv \nu Z.\, \Psi^c \wedge (\Phi^c \vee [-]Z) \end{aligned}$$

The fixed point versions of these formulas are arguably easier to understand than their CTL versions. For instance, the second formula expresses "Ψ^c unless $\Phi^c \wedge \Psi^c$ is true."

A formula that contains free occurrences of variables does not have an explicit complement. For example, Z does not have a complement in μM. However, the semantics of free formulas require valuations. Therefore, we can appeal to the complement valuation V^c, as in Section 4.6. Consequently, $E \models_{\mathsf{V}} \Phi$ iff $E \not\models_{\mathsf{V}^c} \Phi^c$.

Bound variables can be changed in formulas without affecting their meaning. If Y does not occur at all in $\sigma Z.\, \Phi$, then this formula can be rewritten $\sigma Y.\, (\Phi\{Y/Z\})$. It is useful to write formulas in such a way that bound variables be unique. This supports the following definition.

Definition 1 A formula Φ is "normal" provided that

1. if $\sigma_1 Z_1$ and $\sigma_2 Z_2$ are two different occurrences of binders in Φ then $Z_1 \neq Z_2$; and
2. no occurrence of a free variable Z is also used in a binder σZ in Φ.

Every formula can be easily converted into a normal formula of the same size and shape by renaming bound variables.

$$\mu Z. \langle J \rangle Y \vee (\langle b \rangle Z \wedge \mu Y. \nu Z. ([b]Y \wedge [K]Z))$$

can be rewritten as

$$\mu Z. \langle J \rangle Y \vee (\langle b \rangle Z \wedge \mu X. \nu U. ([b]X \wedge [K]U)).$$

This helps us to understand which occurrences of variables are free, and which occurrences are bound. In the sequel, we shall exclusively make use of normal formulas.

In later applications, we shall need to know the set of subformulas of a formula Φ. This set $\mathrm{Sub}(\Phi)$ is finite and extends the definition provided in Section 2.2 for modal formulas.

Definition 2 $\mathrm{Sub}(\Phi)$ is defined inductively by case analysis on Φ

$$\begin{array}{lcl}
\mathrm{Sub}(\mathtt{tt}) & = & \{\mathtt{tt}\} \\
\mathrm{Sub}(\mathtt{ff}) & = & \{\mathtt{ff}\} \\
\mathrm{Sub}(X) & = & \{X\} \\
\mathrm{Sub}(\Phi_1 \wedge \Phi_2) & = & \{\Phi_1 \wedge \Phi_2\} \cup \mathrm{Sub}(\Phi_1) \cup \mathrm{Sub}(\Phi_2) \\
\mathrm{Sub}(\Phi_1 \vee \Phi_2) & = & \{\Phi_1 \vee \Phi_2\} \cup \mathrm{Sub}(\Phi_1) \cup \mathrm{Sub}(\Phi_2) \\
\mathrm{Sub}([K]\Phi) & = & \{[K]\Phi\} \cup \mathrm{Sub}(\Phi) \\
\mathrm{Sub}(\langle K \rangle \Phi) & = & \{\langle K \rangle \Phi\} \cup \mathrm{Sub}(\Phi) \\
\mathrm{Sub}(\nu Z. \Phi) & = & \{\nu Z. \Phi\} \cup \mathrm{Sub}(\Phi) \\
\mathrm{Sub}(\mu Z. \Phi) & = & \{\mu Z. \Phi\} \cup \mathrm{Sub}(\Phi)
\end{array}$$

Example 1 $\mathrm{Sub}(\mu X. \nu Y. ([b]X \wedge [K]Y))$ is the set

$$\{\mu X. \nu Y. ([b]X \wedge [K]Y), \nu Y. [b]X \wedge [K]Y, [b]X \wedge [K]Y, [b]X, [K]Y, X, Y\},$$

which contains seven subformulas.

If Φ is normal and $\sigma Z. \Psi$ belongs to $\mathrm{Sub}(\Phi)$, then the binding variable Z can be used to uniquely identify this subformula.

Later we shall need to understand when one fixed point formula is more "outermost" than another. For this we introduce the notion of subsumption.

Definition 3 Assume Φ is normal and that $\sigma_1 X.\Psi, \sigma_2 Z.\Psi' \in \mathrm{Sub}(\Phi)$. The variable X "subsumes" Z if $\sigma_2 Z.\Psi' \in \mathrm{Sub}(\sigma_1 X.\Psi)$.

For instance, in the case of the formula of Example 1, X subsumes Y but not vice versa, since $\nu Y. ([b]X \wedge [K]Y) \in \mathrm{Sub}(\mu X. \nu Y. ([b]X \wedge [K]Y))$. The following are some simple but useful properties of subsumption whose proofs are left as an exercise for the reader.

Proposition 1

1. *X subsumes X.*
2. *If X subsumes Z and Z subsumes Y, then X subsumes Y.*
3. *If X subsumes Y and $X \neq Y$, then not Y subsumes X.*

Exercises

1. Introduce the following operators of CTL as macros by defining them as fixed points AF, EF, AG and EG.
2. Put the following formulas into normal form
   a. $\mu Y. [K]Y \wedge \nu Y. \langle a \rangle Y$
   b. $[J]Y \wedge \langle K \rangle Y$
   c. $\nu X. \mu Y. (Z \vee \mu Z. \nu X. \langle b \rangle (X \wedge Z) \vee [a]Y)$
3. A normal μM formula is "singular" if each occurrence of a binder σZ binds exactly one occurrence of Z. Prove that every formula can be rewritten as a singular formula. (Hint: show that $\sigma Z.\Phi(Z, Z)$ is equivalent to $\sigma Z_1.\sigma Z_2.\Phi(Z_1, Z_2)$.)
4. Work out the following.
   a. $\text{Sub}(\mathtt{tt} \wedge (\mathtt{tt} \vee \langle a \rangle \mathtt{ff}))$
   b. $\text{Sub}(\nu X. \mu Y. [K]X \vee (\langle - \rangle X \wedge [-K][-]Y))$
   c. $\text{Sub}(\mu X. \langle a \rangle X \wedge \langle a \rangle \langle a \rangle X)$
   d. $\text{Sub}(\mu Y. [b]\mathtt{ff} \wedge [a](\mu Z. [a]\mathtt{ff} \wedge [b]Y \wedge [-b]Z) \wedge [-a]Y)$
   e. $\text{Sub}(\nu Z. [a](\mu Y. \langle - \rangle \mathtt{tt} \wedge [-b]Y) \wedge [-]Z)$
5. For each of the following, determine whether X subsumes Y.
   a. $\nu X. \mu Y. [K]X \vee (\langle - \rangle X \wedge [-K]Y)$
   b. $\mu Y. \nu X. \langle a \rangle X \wedge \langle a \rangle \langle a \rangle X$
   c. $\nu Z. ((\nu X. \langle a \rangle X) \vee (\mu Y. [b]Y)) \wedge [-]Z$
   d. $\mu X. \nu Y. [a]X \wedge [-a]Y$
6. Prove Proposition 1.

5.3 Observable modal logic with fixed points

In Sections 2.4 and 2.5, the observable modal logics M^o and $\text{M}^{o\downarrow}$ are described. Their modal operators such as $\langle\langle\ \rangle\rangle$, $[\![\ ]\!]$, $[\![\downarrow]\!]$ and $\langle\langle\uparrow\rangle\rangle$ are not definable in the modal logic M. However, they are definable in μM, as follows (where as usual it is assumed that Z is not free in Φ).

$\langle\langle\ \rangle\rangle\Phi$	$\stackrel{\text{def}}{=}$	$\mu Z.\,\Phi \vee \langle\tau\rangle Z$
$[\![\ ]\!]\,\Phi$	$\stackrel{\text{def}}{=}$	$\nu Z.\,\Phi \wedge [\tau]Z$
$\langle\langle\uparrow\rangle\rangle\Phi$	$\stackrel{\text{def}}{=}$	$\nu Z.\,\Phi \vee \langle\tau\rangle Z$
$[\![\downarrow]\!]\,\Phi$	$\stackrel{\text{def}}{=}$	$\mu Z.\,\Phi \wedge [\tau]Z$

The contrast in meaning between $[\![\ ]\!]$ and $[\![\downarrow]\!]$ is the difference in their fixed point. The formula $[\![\ ]\!]\,\Phi$ is just $[\tau^*]\Phi$, since Φ has to hold throughout any amount of silent activity. A divergent process may have the property $[\![\ ]\!]\,\Phi$, but it cannot satisfy $[\![\downarrow]\!]\,\Phi$. Hence, the change in fixed point. The set

$$\bigcap\{\mathsf{E} \subseteq \mathsf{P} \;:\; \|\,\Phi \wedge [\tau]Z\,\|^{\mathsf{P}}_{\mathsf{V}[\mathsf{E}/Z]} \subseteq \mathsf{E}\}$$

at least contains stable processes (those unable to perform a silent action) with the property Φ. Therefore, the set also contains any process that satisfies Φ and that eventually stabilizes after some amount of silent activity, provided that Φ continues to be true.

The modal logics M^o and $\mathrm{M}^{o\downarrow}$ are therefore special sublogics of $\mu\mathrm{M}$. The derived operators $[\![K]\!]$ and $[\![\downarrow K]\!]$ are defined using embedded fixed points, as follows (where Φ does not contain Z or Y).

$$\begin{aligned}
[\![K]\!]\,\Phi &\stackrel{\text{def}}{=} [\![\ ]\!]\,[K]\,[\![\ ]\!]\,\Phi \\
&= \nu Z.\,[K]\,[\![\ ]\!]\,\Phi \wedge [\tau]Z \\
&= \nu Z.\,[K](\nu Y.\,\Phi \wedge [\tau]Y) \wedge [\tau]Z \\
[\![\downarrow K]\!]\,\Phi &\stackrel{\text{def}}{=} [\![\downarrow]\!]\,[K]\,[\![\ ]\!]\,\Phi \\
&= \mu Z.\,[K]\,[\![\ ]\!]\,\Phi \wedge [\tau]Z \\
&= \mu Z.\,[K](\nu Y.\,\Phi \wedge [\tau]Y) \wedge [\tau]Z
\end{aligned}$$

Observable modal logic M^o can be extended with fixed points. The formulas of observable mu-calculus, $\mu\mathrm{M}^o$, are as follows.

$$\begin{aligned}
\Phi \;::=\;& Z \mid \mathtt{tt} \mid \mathtt{ff} \mid \Phi_1 \wedge \Phi_2 \mid \Phi_1 \vee \Phi_2 \mid [\![K]\!]\,\Phi \mid \langle\langle K\rangle\rangle\Phi \mid \\
& [\![\ ]\!]\,\Phi \mid \langle\langle\ \rangle\rangle\Phi \mid \nu Z.\,\Phi \mid \mu Z.\,\Phi
\end{aligned}$$

Here K ranges over sets of observable actions (which exclude τ). This sublogic is suitable for describing properties of observable transition systems.

We can also define the sublogic $\mu\mathrm{M}^{o\downarrow}$, which contains the additional modalities $[\![\downarrow]\!]$ and $\langle\langle\uparrow\rangle\rangle$. Unlike $\mu\mathrm{M}^o$, this logic is sensitive to divergence.

Exercises

1. Show that the fixed point definitions of $\langle\langle\ \rangle\rangle$, $[\![\ ]\!]$, $\langle\langle\uparrow\rangle\rangle$ and $[\![\downarrow]\!]$ are indeed correct.

2. Provide fixed point definitions for the following modalities $[\![K\downarrow]\!]$, $[\![\downarrow K\downarrow]\!]$, $\langle\!\langle K\rangle\!\rangle$, and $\langle\!\langle\uparrow K\uparrow\rangle\!\rangle$.
3. Show that divergence is not definable in μM^o. That is, prove that there is not a formula $\Phi \in \mu M^o$ with the feature that $E\uparrow$ iff $E \models \Phi$ for any E.
4. Show that `Protocol` and `Cop` have the same μM^o properties, but not the same $\mu M^{o\downarrow}$ properties.

5.4 Preservation of bisimulation equivalence

The modal logic M characterizes strong bisimulation equivalence, as shown in Section 3.4. There are two parts to characterisation.

1. Two bisimilar processes have the same modal properties
2. Two image-finite processes having the same modal properties are bisimilar.

Because M is a sublogic of μM, 2 is also true for μM (and in fact for any extension of modal logic). Let Γ be the set of formulas of μM which do not contain free variables. Recall that $E \equiv_\Gamma F$ abbreviates that E and F share the same Γ properties.

Proposition 1 *If E and F are image-finite, and $E \equiv_\Gamma F$, then $E \sim F$.*

The proof of Proposition 1 does not rely on the extra expressive power of μM over and above M. However, one may ask whether the restriction to image-finite processes is still essential to this result given that fixed points are expressible using infinitary conjunction and disjunction, and that infinitary modal logic M_∞ characterizes bisimulation equivalence exactly. In Section 3.4 two examples of clocks showed that image finiteness is essential in the case of modal formulas. However, although these clocks have the same modal properties, they do not have the same μM properties. One of the clocks has an infinite tick capability, expressed by the formula $\nu Z.\langle\mathtt{tick}\rangle$, that the other fails to satisfy. The following example, due to Roope Kaivola, shows that image finiteness (or a weakened version of it) is still necessary.

Example 1 Let $\{\mathtt{Q}_i : i \in I\}$ be the set of all finite state processes whose actions belong to $\{a, b\}$, and assuming $n \in \mathbb{N}$, consider the following processes.

$$\begin{aligned}
\mathtt{P}(n) &\stackrel{\text{def}}{=} a^n.b.\mathtt{P}(n+1)\\
\mathtt{R} &\stackrel{\text{def}}{=} \sum\{a.\mathtt{Q}_i : i \in I\}\\
\mathtt{P} &\stackrel{\text{def}}{=} \mathtt{P}(1) + \mathtt{R}
\end{aligned}$$

The behaviour of $\mathtt{P}(1)$ is as follows.

$$\mathtt{P}(1) \stackrel{ab}{\longrightarrow} \mathtt{P}(2) \stackrel{aab}{\longrightarrow} \mathtt{P}(3) \stackrel{aaab}{\longrightarrow} \ldots \stackrel{a^n b}{\longrightarrow} \mathtt{P}(n+1) \stackrel{a^{n+1}b}{\longrightarrow} \ldots$$

The processes P and R are not bisimilar because no finite state process can be bisimulation equivalent to $b.\mathtt{P}(2)$ (via the pumping lemma for regular languages). However, P and R have the same μM properties, when expressed by formulas without free variables. To see this, suppose that there is a formula Φ that distinguishes between P and R: it follows that there is a formula Ψ such that $b.\mathtt{P}(2) \models \Psi$ and $\mathtt{Q}_i \not\models \Psi$ for all $i \in I$. By the finite model theorem, Proposition 3 of Section 5.1, there is a finite state process E such that $E \models \Psi$. A small argument shows that E can be built from the actions a and b. Consequently, E is $\mathtt{Q}_j$ for some $j \in I$. But this contradicts that every process $\mathtt{Q}_i$ fails to have the property Ψ.

Not only do bisimilar processes have the same modal properties, but they also have the same μM properties.

Proposition 2 *If $E \sim F$, then $E \equiv_\Gamma F$.*

An indirect proof of this Proposition uses the facts that it holds for the logic M_∞, and that μM is a sublogic of M_∞, as we shall see in the next section. However, we shall prove this result directly; for we wish to expose some of the inductive structure of modal mu-calculus.

A subset E of P is bisimulation closed if, whenever $E \in \mathsf{E}$ and $F \in \mathsf{P}$ and $E \sim F$, then F is also in E. The desired result, Proposition 2, is equivalent to the claim that, for any formula Φ without free variables and set of processes P, the set $\| \Phi \|^{\mathsf{P}}$ is bisimulation closed. If this is true, then it is not possible that there can be a μM formula Φ and a pair of bisimilar processes E and F such that $E \models \Phi$ and $F \not\models \Phi$. Conversely, if $\| \Phi \|^{\mathsf{P}}$ is bisimulation closed, then any pair of processes E and F such that $E \models \Phi$ and $F \not\models \Phi$ can not be bisimilar. The next lemma states some straightforward features of bisimulation closure.

Lemma 1 *If E and F are bisimulation closed subsets of P, then*

1. $\mathsf{E} \cap \mathsf{F}$ *and* $\mathsf{E} \cup \mathsf{F}$ *are bisimulation closed,*
2. $\{E \in \mathsf{P} : \text{if } E \stackrel{a}{\longrightarrow} F \text{ and } a \in K \text{ then } F \in \mathsf{E}\}$ *is bisimulation closed,*
3. $\{E \in \mathsf{P} : \exists F \in \mathsf{E}. \exists a \in K.\ E \stackrel{a}{\longrightarrow} F\}$ *is bisimulation closed.*

Proof. Assume that the subsets E and F of P are bisimulation closed. If $E \in \mathsf{E} \cap \mathsf{F}$, then $E \in \mathsf{E}$ and $E \in \mathsf{F}$. Consequently, if $E \sim F$ and $F \in \mathsf{P}$, then $F \in \mathsf{E}$ and $F \in \mathsf{F}$, and so $F \in \mathsf{E} \cap \mathsf{F}$. Assume $E \in \mathsf{E} \cup \mathsf{F}$. Therefore $E \in \mathsf{E}$ or $E \in \mathsf{F}$. If $E \sim F$ and $F \in \mathsf{P}$, then $F \in \mathsf{E}$ or $F \in \mathsf{F}$ because these sets are bisimulation closed, and therefore $\mathsf{E} \cup \mathsf{F}$ is also bisimulation closed. For 2, suppose E belongs to $\mathsf{G} = \{G \in \mathsf{P} : \text{if } G \stackrel{a}{\longrightarrow} H \text{ and } a \in K \text{ then } H \in \mathsf{E}\}$ and $E \sim F$ with $F \in \mathsf{P}$. To show that F is also in G, it suffices to demonstrate that, if $F \stackrel{a}{\longrightarrow} F'$ when $a \in K$, then $F' \in \mathsf{E}$. Because P is transition closed, $F' \in \mathsf{P}$. Moreover, because $E \sim F$ it follows that $E \stackrel{a}{\longrightarrow} E'$ and $E' \sim F'$, for some $E' \in \mathsf{E}$. And therefore $F' \in \mathsf{E}$ because E is bisimulation closed. Part 3 has a similar proof. □

Associated with any subset E of P are the following two subsets.

$$\mathsf{E}^d = \{E \in \mathsf{E} : \text{ if } E \sim F \text{ and } F \in \mathsf{P} \text{ then } F \in \mathsf{E}\}$$
$$\mathsf{E}^u = \{E \in \mathsf{P} : \exists F \in \mathsf{E}.\ E \sim F\}$$

The set E^d is the *largest* bisimulation closed subset of E, and E^u is the *smallest* bisimulation closed superset of E (both with respect to P).

Lemma 2 *For any subsets* E *and* F *of* P

1. E^d *and* E^u *are bisimulation closed,*
2. $\mathsf{E}^d \subseteq \mathsf{E} \subseteq \mathsf{E}^u$,
3. *if* E *is bisimulation closed, then* $\mathsf{E}^d = \mathsf{E}^u$,
4. *if* $\mathsf{E} \subseteq \mathsf{F}$, *then* $\mathsf{E}^d \subseteq \mathsf{F}^d$ *and* $\mathsf{E}^u \subseteq \mathsf{F}^u$.

Proof. These are straightforward consequences of the definitions of E^d and E^u. □

A valuation V is bisimulation closed if, for each variable Z, the set $\mathsf{V}(Z)$ is bisimulation closed. Therefore, we can associate the bisimulation closed valuations V^d and V^u with any valuation V: for any variable Z, the set $\mathsf{V}^d(Z) = (\mathsf{V}(Z))^d$ and $\mathsf{V}^u(Z) = (\mathsf{V}(Z))^u$. Proposition 2 is a corollary of the following result, in which Φ is an arbitrary formula of modal mu-calculus and therefore may contain free variables.

Proposition 3 *If* V *is bisimulation closed, then* $\| \Phi \|_{\mathsf{V}}^{\mathsf{P}}$ *is bisimulation closed.*

Proof. The proof proceeds by simultaneous induction on the structure of Φ with the following three propositions.

1. If V is bisimulation closed, then $\| \Phi \|_{\mathsf{V}}^{\mathsf{P}}$ is bisimulation closed
2. If $\| \Phi \|_{\mathsf{V}}^{\mathsf{P}} \subseteq \mathsf{E}$, then $\| \Phi \|_{\mathsf{V}^d}^{\mathsf{P}} \subseteq \mathsf{E}^d$
3. If $\mathsf{E} \subseteq \| \Phi \|_{\mathsf{V}}^{\mathsf{P}}$, then $\mathsf{E}^u \subseteq \| \Phi \|_{\mathsf{V}^u}^{\mathsf{P}}$

We now drop the index P. The base cases are when Φ is $\mathtt{tt}$, $\mathtt{ff}$ or a variable Z. The first two are clear. Suppose Φ is Z. Because V is bisimulation closed, it follows that $\| Z \|_{\mathsf{V}}$ is also bisimulation closed. For 2, suppose $\mathsf{V}(Z) \subseteq \mathsf{E}$. By Lemma 2 part 4, it follows that $\mathsf{V}^d(Z) \subseteq \mathsf{E}^d$. Similarly for 3, if $\mathsf{E} \subseteq \mathsf{V}(Z)$, then $\mathsf{E}^u \subseteq \mathsf{V}^u(Z)$. The induction step divides into the various subcases. Suppose $\Phi = \Phi_1 \wedge \Phi_2$. For 1, since $\| \Phi \|_{\mathsf{V}} = \| \Phi_1 \|_{\mathsf{V}} \cap \| \Phi_2 \|_{\mathsf{V}}$ and by the induction hypothesis both $\| \Phi_i \|_{\mathsf{V}}$ are bisimulation closed, it follows from Lemma 1 part 1 that $\| \Phi \|_{\mathsf{V}}$ is also bisimulation closed. A similar argument establishes that $\| \Phi \|_{\mathsf{V}^d}$ is bisimulation closed. By monotonicity, $\| \Phi \|_{\mathsf{V}^d} \subseteq \| \Phi \|_{\mathsf{V}}$ and so $\| \Phi \|_{\mathsf{V}^d} \subseteq \mathsf{E}$, and therefore $\| \Phi \|_{\mathsf{V}^d} \subseteq \mathsf{E}^d$ as required for 2. For 3, assume that $E \in \mathsf{E}^u$. By the definition of E^u, there is an $F \in \mathsf{E}$ such that $F \sim E$. But then $F \in \| \Phi \|_{\mathsf{V}}$ and therefore, by monotonicity, $F \in \| \Phi \|_{\mathsf{V}^u}$. The same argument as in case 1 shows that $\| \Phi \|_{\mathsf{V}^u}$ is bisimulation closed, and therefore $E \in \| \Phi \|_{\mathsf{V}^u}$. The case of $\Phi = \Phi_1 \vee \Phi_2$ is similar. Suppose $\Phi = [K]\Psi$. For 1, by the induction hypothesis $\| \Psi \|_{\mathsf{V}}$ is bisimulation closed, and therefore by Lemma 1 part 2 the set $\| \Phi \|_{\mathsf{V}}$ is as well. The arguments for 2 and 3 follow those for the case of $\wedge$, above: both sets $\| \Phi \|_{\mathsf{V}^d}$, $\| \Phi \|_{\mathsf{V}^u}$ are bisimulation closed. So, if $E \in \| \Phi \|_{\mathsf{V}^d}$, then $E \in \| \Phi \|_{\mathsf{V}}$ and also $E \in \mathsf{E}^d$. And if $E \in \mathsf{E}^u$, then there is some $F \in \mathsf{E}$ such that $F \sim E$, meaning $F \in \| \Phi \|_{\mathsf{V}}$ and therefore $F \in \| \Phi \|_{\mathsf{V}^u}$. The other modal case when $\Phi = \langle K \rangle \Psi$ is similar, using Lemma 1 part 3.

The interesting cases are when Φ is a fixed point formula. Suppose first that Φ is $\mu Z. \Psi$. To show 1, we need to establish that the least E such that $\| \Psi \|_{\mathsf{V}[\mathsf{E}/Z]} \subseteq \mathsf{E}$ is bisimulation closed. By assumption, V is bisimulation closed, and so $\mathsf{V} = \mathsf{V}^d$ from Lemma 2 parts 2 and 3. Therefore, by the induction hypothesis on 2, we know that $\| \Psi \|_{\mathsf{V}[\mathsf{E}^d/Z]} \subseteq \mathsf{E}^d$. Because $\mathsf{E}^d \subseteq \mathsf{E}$ and E is the least set obeying $\| \Psi \|_{\mathsf{V}[\mathsf{E}/Z]} \subseteq \mathsf{E}$, it follows that $\mathsf{E} = \mathsf{E}^d$ and is therefore bisimulation closed. Cases 2 and 3 follow the same pattern as before. The final case is $\Phi = \nu Z. \Psi$. The argument is similar to that just employed for the least fixed point, except that to establish 1 we use the induction hypothesis on 3: we need to show that the largest set E such that $\mathsf{E} \subseteq \| \Psi \|_{\mathsf{V}[\mathsf{E}/Z]}$ is bisimulation closed. Because V is bisimulation closed $\mathsf{V} = \mathsf{V}^u$, so $\mathsf{E}^u \subseteq \| \Psi \|_{\mathsf{V}[\mathsf{E}^u/Z]}$ by the induction hypothesis on 3, and therefore $\mathsf{E} = \mathsf{E}^u$. Cases 2 and 3 follow as before. □

This result tells us more than that bisimilar processes have the same properties when expressed as a formula without free variables. They also have the same properties when expressed by formulas with free variables, provided the meanings of the free variables are bisimulation closed. The proof of this result also establishes that formulas of $\mu \mathrm{M}^o$, observable modal mu-calculus, which do not contain free variables, are preserved by observable bisimulation equivalence. The earlier lemmas and Proposition 3 all hold for observable bisimulation equivalence, and $\mu \mathrm{M}^o$.

Exercises

1. Prove Lemma 1 part 3.
2. Show that if E is a bisimulation closed subset of P, then its complement $\mathsf{P} - \mathsf{E}$ is also bisimulation closed.
3. Let $\{\mathsf{E}_i : i \in I\}$ be an indexed family of bisimulation closed subsets of P. Show that $\bigcap \{\mathsf{E}_i : i \in I\}$ and $\bigcup \{\mathsf{E}_i : i \in I\}$ are bisimulation closed.
4. Prove Lemma 2.
5. Extend modal mu-calculus so that there is a formula Φ that does not contain free variables such that $\| \Phi \|$ is *not* bisimulation closed.

5.5 Approximants

At first sight, there is a chasm between the meaning of an extremal fixed point and techniques (other than exhaustive analysis) for actually finding it. However, there is a more mechanical method, an iterative technique, due to Tarski and others, for discovering least and greatest fixed points. Let μg be the least fixed point, and νg the greatest fixed point, of the monotonic function $g : 2^P \rightarrow 2^P$. From Proposition 1 of Section 4.5 the following hold.

$$\begin{array}{lcl} \mu g & = & \bigcap\{\mathsf{E} \subseteq \mathsf{P} : g(\mathsf{E}) \subseteq \mathsf{E}\} \\ \nu g & = & \bigcup\{\mathsf{E} \subseteq \mathsf{P} : \mathsf{E} \subseteq g(\mathsf{E})\} \end{array}$$

Suppose we wish to determine the set νg. Let $\nu^i g$ for $i \geq 0$ be defined iteratively as follows.

$$\nu^0 g = \mathsf{P} \qquad \nu^{i+1} g = g(\nu^i g)$$

Because P is the largest subset of itself, $\nu^1 g \subseteq \nu^0 g$ and, by monotonicity of g, this implies that $g(\nu^1 g) \subseteq g(\nu^0 g)$, that is $\nu^2 g \subseteq \nu^1 g$. Applying g again to both sides, $\nu^3 g \subseteq \nu^2 g$, and consequently for all i, $\nu^{i+1} g \subseteq \nu^i g$. Moreover, the required fixed point set νg is a subset of every $\nu^i g$. First, $\nu g \subseteq \nu^0 g$ and so by monotonicity $g(\nu g) \subseteq \nu^1 g$. Because νg is a fixed point of g, $g(\nu g) = \nu g$, and therefore $\nu g \subseteq \nu^1 g$. Using montonicity of g once more we obtain $\nu g \subseteq \nu^2 g$. Consequently, with repeated application of g it follows that $\nu g \subseteq \nu^i g$ for any i. Therefore, we have the following situation.

$$\begin{array}{ccccccccc} \nu^0 g & \supseteq & \nu^1 g & \supseteq & \dots & \supseteq & \nu^i g & \supseteq & \dots \\ \cup & & \cup & & & & \cup & & \\ \nu g & & \nu g & & \dots & & \nu g & & \dots \end{array}$$

If $\nu^i g = \nu^{i+1} g$, then the fixed point νg is $\nu^i g$. Why is this? Because $\nu^i g$ is then a fixed point of g, and $\nu g \subseteq \nu^i g$ and νg is the greatest fixed point.

These observations suggest a strategy for discovering νg. Iteratively construct the sets $\nu^i g$ starting with $i = 0$, until $\nu^i g$ is the same set as $\nu^{i+1} g$. If P is a finite set containing n processes, then this iterative construction terminates at, or before, the case $i = n$, and therefore νg is equal to $\nu^n g$.

Example 1 Let P be $\{\mathtt{Cl}, \mathtt{tick.0}, 0\}$, and let g be the following function $f[\langle\mathtt{tick}\rangle Z, Z]$.

$$\begin{array}{lclcl} \nu^0 g & = & \mathsf{P} & = & \{\mathtt{Cl}, \mathtt{tick.0}, 0\} \\ \nu^1 g & = & \| \langle\mathtt{tick}\rangle Z \|^{\mathsf{P}}_{\mathsf{V}[\nu^0 g/Z]} & = & \{\mathtt{Cl}, \mathtt{tick.0}\} \\ \nu^2 g & = & \| \langle\mathtt{tick}\rangle Z \|^{\mathsf{P}}_{\mathsf{V}[\nu^1 g/Z]} & = & \{\mathtt{Cl}\} \\ \nu^3 g & = & \| \langle\mathtt{tick}\rangle Z \|^{\mathsf{P}}_{\mathsf{V}[\nu^2 g/Z]} & = & \{\mathtt{Cl}\} \end{array}$$

Stabilization occurs at the stage $\nu^2 g$ because this set coincides with $\nu^3 g$. The fixed point νg, the set of processes $\| \nu Z. \langle\mathtt{tick}\rangle Z \|^{\mathsf{P}}_{\mathsf{V}}$, is therefore the singleton set $\{\mathtt{Cl}\}$.

If P is not a finite set of processes, then we can still guarantee that νg is reachable iteratively by invoking ordinals as indices. Recall that ordinals are ordered as follows.

$$0, 1, \ldots, \omega, \omega + 1, \ldots, \omega + \omega, \omega + \omega + 1, \ldots$$

The ordinal ω is the initial limit ordinal (which has no immediate predecessor), whereas $\omega + 1$ is its successor. Assume that α and λ range over ordinals. We define $\nu^{\alpha} g$, for any ordinal $\alpha \geq 0$, with the base case and successor case as before, $\nu^0 g = \mathsf{P}$ and $\nu^{\alpha+1} g = g(\nu^{\alpha} g)$. The case when λ is a limit ordinal is defined as follows.

$$\nu^{\lambda} g \;=\; \bigcap \{\nu^{\alpha} g \;:\; \alpha < \lambda\}$$

By the same arguments as above there is the following possibly decreasing sequence[3].

$$\begin{array}{ccccccccc} \nu^0 g & \supseteq & \ldots & \supseteq & \nu^{\omega} g & \supseteq & \nu^{\omega+1} g & \supseteq & \ldots \\ \cup & & & & \cup & & \cup & & \\ \nu g & & \ldots & & \nu g & & \nu g & & \ldots \end{array}$$

The fixed point set νg appears somewhere in the sequence, at the first point when $\nu^{\alpha} g = \nu^{\alpha+1} g$.

Example 2 Let P be the set $\{\mathtt{C}, \mathtt{B}_i \;:\; i \geq 0\}$ when $\mathtt{C}$ is the cell

$$\begin{array}{lcll} \mathtt{C} & \stackrel{\text{def}}{=} & \mathtt{in}(x).\mathtt{B}_x & \text{where } x : \mathbb{N} \\ \mathtt{B}_{n+1} & \stackrel{\text{def}}{=} & \mathtt{down}.\mathtt{B}_n & \text{for } n \geq 0 \end{array}$$

Let g be the function $f[\langle - \rangle Z, Z]$. The fixed point νg is the empty set.

$$\begin{array}{lclcl} \nu^0 g & = & \mathsf{P} & = & \{\mathtt{C}, \mathtt{B}_i \;:\; i \geq 0\} \\ \nu^1 g & = & \| \langle - \rangle Z \|^{\mathsf{P}}_{\mathsf{V}[\nu^0 g/Z]} & = & \{\mathtt{C}, \mathtt{B}_i \;:\; i \geq 1\} \\ \vdots & & \vdots & & \\ \nu^{j+1} g & = & \| \langle - \rangle Z \|^{\mathsf{P}}_{\mathsf{V}[\nu^j g/Z]} & = & \{\mathtt{C}, \mathtt{B}_i \;:\; i \geq j+1\} \end{array}$$

The set $\nu^{\omega} g$ is $\bigcap \{\nu^i g \;:\; i < \omega\}$, which is $\{\mathtt{C}\}$ because each $\mathtt{B}_j$ is excluded from $\nu^{j+1} g$. The very next iterate is the fixed point, $\nu^{\omega+1} g$ is $\| \langle - \rangle Z \|^{\mathsf{P}}_{\mathsf{V}[\nu^{\omega} g/Z]} = \emptyset$. So, stabilization occurs at $\nu^{\omega+1} g$. This example can be further extended.

$$\mathtt{C}' \stackrel{\text{def}}{=} \mathtt{in}(x).\mathtt{B}'_x \quad \mathtt{B}'_{n+2} \stackrel{\text{def}}{=} \mathtt{down}.\mathtt{B}'_{n+1} \quad \mathtt{B}'_1 \stackrel{\text{def}}{=} \mathtt{down}.\mathtt{C}$$

[3]Notice that $\nu g \subseteq \nu^{\lambda} g$ when λ is a limit ordinal because $\nu g \subseteq \nu^{\alpha} g$ for all $\alpha < \lambda$.

Consider the following iterates.

$$\begin{array}{lcl} \nu^i g & = & \{\mathtt{C}', \mathtt{B}'_k, \mathtt{C}, \mathtt{B}_j \ : \ k \geq 1 \text{ and } j \geq i\} \\ \vdots & & \vdots \\ \nu^{\omega+1} g & = & \{\mathtt{C}', \mathtt{B}'_k \ : \ k \geq 1\} \\ \vdots & & \vdots \\ \nu^{\omega+\omega} g & = & \{\mathtt{C}'\} \\ \nu^{\omega+\omega+1} g & = & \emptyset \end{array}$$

The fixed point νg therefore stabilizes at stage $\omega + \omega + 1$.

The situation for a least fixed point μg is dual. Let $\mu^0 g$ be the smallest subset of P, that is $\emptyset$, and let $\mu^{\alpha+1} g = g(\mu^\alpha g)$. The following defines a limit ordinal λ.

$$\mu^\lambda g \ = \ \bigcup \{\mu^\alpha g \ : \ \alpha < \lambda\}$$

There is the following possibly increasing sequence of sets.

$$\begin{array}{ccccccccc} \mu g & & \ldots & & \mu g & & \mu g & & \ldots \\ \cup & & & & \cup & & \cup & & \\ \mu^0 g & \subseteq & \ldots & \subseteq & \mu^\omega g & \subseteq & \mu^{\omega+1} g & \subseteq & \ldots \end{array}$$

The fixed point μg is a *superset* of each of the iterates $\mu^\alpha g$. First, $\mu^0 g \subseteq \mu g$, and so by monotonicity of g (and limit considerations) $\mu^\alpha g \subseteq \mu g$ for any α. The first point that $\mu^\alpha g$ is equal to its successor $\mu^{\alpha+1} g$ is the required fixed point μg. An iterative method for finding μg is to construct the sets $\mu^\alpha g$ starting with $\mu^0 g$ until it is the same as its successor.

Example 3 Let g be be the function $f[[\mathtt{tick}]\mathtt{ff} \vee \langle - \rangle Z, Z]$. Assume P is the set of processes in example 1 above.

$$\begin{array}{lclcl} \mu^0 g & = & \emptyset & & \\ \mu^1 g & = & \| \, [\mathtt{tick}]\mathtt{ff} \vee \langle - \rangle Z \, \|^{\mathsf{P}}_{\mathsf{V}[\mu^0 g/Z]} & = & \{\mathtt{0}\} \\ \mu^2 g & = & \| \, [\mathtt{tick}]\mathtt{ff} \vee \langle - \rangle Z \, \|^{\mathsf{P}}_{\mathsf{V}[\mu^1 g/Z]} & = & \{\mathtt{tick.0}, \mathtt{0}\} \\ \mu^3 g & = & \| \, [\mathtt{tick}]\mathtt{ff} \vee \langle - \rangle Z \, \|^{\mathsf{P}}_{\mathsf{V}[\mu^2 g/Z]} & = & \{\mathtt{tick.0}, \mathtt{0}\} \end{array}$$

Stabilization occurs at $\mu^2 g$, which is the required fixed point. Notice that if we consider νg instead, then we obtain the following different set.

$$\begin{array}{lclcl} \nu^0 g & = & \mathsf{P} & = & \{\mathtt{Cl}, \mathtt{tick.0}, \mathtt{0}\} \\ \nu^1 g & = & \| \, [\mathtt{tick}]\mathtt{ff} \vee \langle - \rangle Z \, \|^{\mathsf{P}}_{\mathsf{V}[\nu^0 g/Z]} & = & \mathsf{P} \end{array}$$

This stabilizes at the initial point.

Example 4 Consider the following family of clocks, $\mathtt{Cl}^i$, $i > 0$.

$$\mathtt{Cl}^1 \stackrel{\text{def}}{=} \mathtt{tick.0}$$
$$\mathtt{Cl}^{i+1} \stackrel{\text{def}}{=} \mathtt{tick.Cl}^i \quad i \geq 1$$

Let E be $\sum\{\mathtt{Cl}^i : i \geq 1\}$. E models an arbitrary new clock, which will eventually break down. Let P be the set $\{E, \mathtt{0}, \mathtt{Cl}^i : i \geq 1\}$. Because all behaviour is finite, each process in P has the property $\mu Z.\,[\mathtt{tick}]Z$. Let g be the function $f[[\mathtt{tick}]Z, Z]$.

$$\begin{array}{lllll} \mu^0 g & = & \emptyset & & \\ \mu^1 g & = & \| [\mathtt{tick}]Z \|^{\mathsf{P}}_{\mathsf{V}[\mu^0 g/Z]} & = & \{\mathtt{0}\} \\ \vdots & & \vdots & & \\ \mu^{i+1} g & = & \| [\mathtt{tick}]Z \|^{\mathsf{P}}_{\mathsf{V}[\mu^i g/Z]} & = & \{\mathtt{0}, \mathtt{Cl}^j : j < i+1\} \end{array}$$

The initial limit point $\mu^\omega g$ is the following set.

$$\bigcup\{\mu^i g : i < \omega\} = \{\mathtt{0}, \mathtt{Cl}^j : j \geq 0\}$$

At the next stage, the required fixed point is reached.

$$\mu^{\omega+1} g = \| [\mathtt{tick}]Z \|^{\mathsf{P}}_{\mathsf{V}[\mu^\omega g/Z]} = \mathsf{P}$$

Therefore, for all $\alpha \geq \omega + 1$, $\mu^\alpha g = \mathsf{P}$.

Each set $\sigma^\alpha g$ is an approximation to the set σg. Increasing the index α provides a closer approximation. Each $\nu^\alpha g$ approximates νg from above, whereas each $\mu^\alpha g$ approximates μg from below. Consequently, an extremal fixed point is the limit of a sequence of approximants.

There is a more syntactic characterization of the extremal fixed points using the extended modal logic M_∞ of Section 2.2 (which contains infinitary conjunction $\bigwedge$ and disjunction $\bigvee$). If g is the function $f[\Phi, Z]$, then with respect to P and V the element νg is $\| \nu Z.\,\Phi \|^{\mathsf{P}}_{\mathsf{V}}$ and μg is $\| \mu Z.\,\Phi \|^{\mathsf{P}}_{\mathsf{V}}$. The initial approximant $\nu^0 g$ is the set P, which is just $\| \mathtt{tt} \|^{\mathsf{P}}_{\mathsf{V}}$, and the initial approximant $\mu^0 g$ is $\emptyset$, which is $\| \mathtt{ff} \|^{\mathsf{P}}_{\mathsf{V}}$. The element $\nu^1 g$ is $g(\nu^0 g)$, which is $\| \Phi \|^{\mathsf{P}}_{\mathsf{V}[\| \mathtt{tt} \|^{\mathsf{P}}_{\mathsf{V}}/Z]}$, that is $\| \Phi\{\mathtt{tt}/Z\} \|^{\mathsf{P}}_{\mathsf{V}}$. Similarly $\mu^1 g$ is $\| \Phi\{\mathtt{ff}/Z\} \|^{\mathsf{P}}_{\mathsf{V}}$. For each ordinal α, we define $\sigma Z^\alpha.\Phi$ as a formula of the extended modal logic. As before, let λ be a limit ordinal.

$$\begin{array}{lllllll} \nu Z^0.\,\Phi & = & \mathtt{tt} & & \mu Z^0.\,\Phi & = & \mathtt{ff} \\ \nu Z^{\alpha+1}.\,\Phi & = & \Phi\{\nu Z^\alpha.\,\Phi/Z\} & & \mu Z^{\alpha+1}.\,\Phi & = & \Phi\{\mu Z^\alpha.\,\Phi/Z\} \\ \nu Z^\lambda.\,\Phi & = & \bigwedge\{\nu Z^\alpha.\,\Phi : \alpha < \lambda\} & & \mu Z^\lambda.\,\Phi & = & \bigvee\{\mu Z^\alpha.\,\Phi : \alpha < \lambda\} \end{array}$$

Proposition 1 *Fix P and V and let g be the function $f[\Phi, Z]$. Then the approximant $\sigma^\alpha g = \| \sigma Z^\alpha.\,\Phi \|^{\mathsf{P}}_{\mathsf{V}}$ for any ordinal α.*

Proof. By induction on α. We drop the index P. The base cases are straightforward. Suppose the result holds for all $\alpha < \delta$. If δ is a successor ordinal, say $\alpha + 1$, then $\sigma^{\alpha+1}g$ is $g(\sigma^{\alpha}g)$, which by the induction hypothesis is $\| \Phi \|_{\mathsf{V}[\| \sigma Z^{\alpha}.\Phi \|_{\mathsf{V}}/Z]}$, which in turn is equal to $\| \Phi\{\sigma Z^{\alpha}.\Phi/Z\} \|_{\mathsf{V}}$. The result follows because $\Phi\{\sigma Z^{\alpha}.\Phi/Z\}$ is $\sigma Z^{\alpha+1}.\Phi$. If δ is a limit ordinal, then $\nu^{\delta}g$ is $\bigcap\{\nu^{\alpha}g : \alpha < \delta\}$. By the induction hypothesis, this set is $\bigcap\{\| \nu^{\alpha}Z.\Phi \|_{\mathsf{V}} : \alpha < \delta\}$, which is $\| \bigwedge\{\nu^{\alpha}Z.\Phi : \alpha < \delta\} \|_{\mathsf{V}}$, and is therefore $\| \nu^{\delta}Z.\Phi \|_{\mathsf{V}}$. A similar argument establishes the least fixed point case. □

A simple consequence of Proposition 1 is a more direct definition of satisfaction $E \models_{\mathsf{V}} \Phi$ when Φ is a fixed point formula.

$E \models_{\mathsf{V}} \nu Z.\,\Phi$	iff	$E \models \nu Z^{\alpha}.\,\Phi$ for all ordinals α
$E \models_{\mathsf{V}} \mu Z.\,\Phi$	iff	$E \models \mu Z^{\alpha}.\,\Phi$ for some ordinal α

The quantification over ordinals in these clauses is bounded by the size of $\mathsf{P}(E)$. The following is a corollary of the discussion in this section. It will turn out to be very useful later, since it provides "least approximants" for when a process has a least fixed point property, and when a process fails to have a greatest fixed point property.

Proposition 2

1. *If $E \models_{\mathsf{V}} \mu Z.\,\Phi$, then there is a least ordinal α such that $E \models_{\mathsf{V}} \mu Z^{\alpha}.\,\Phi$ and for all $\beta < \alpha$, $E \not\models_{\mathsf{V}} \mu Z^{\beta}.\,\Phi$.*
2. *If $E \not\models_{\mathsf{V}} \nu Z.\,\Phi$, then there is a least ordinal α such that $E \not\models_{\mathsf{V}} \nu Z^{\alpha}.\,\Phi$ and for all $\beta < \alpha$, $E \models_{\mathsf{V}} \nu Z^{\beta}.\,\Phi$.*

Proof. Notice that for any E, $E \not\models_{\mathsf{V}} \mu Z^{0}.\,\Phi$. Therefore, if $E \models_{\mathsf{V}} \mu Z^{\alpha}.\,\Phi$, then for all $\beta > \alpha$ it follows by monotonicity that $E \models_{\mathsf{V}} \mu Z^{\beta}.\,\Phi$. Consequently, there is a least ordinal α for which $E \models_{\mathsf{V}} \mu Z^{\alpha}.\,\Phi$ when E has the property $\mu Z.\,\Phi$ relative to V. Case 2 is dual. □

Example 5 In Section 5.3, the definitions of $[\![\ ]\!]\,\Phi$ and $[\![\downarrow]\!]\,\Phi$ in μM were contrasted. Let $\nu Z.\,\Psi$ be the formula $\nu Z.\,\Phi \wedge [\tau]Z$ (expressing $[\![\ ]\!]\,\Phi$) and let $\mu Z.\,\Psi$ be $\mu Z.\,\Phi \wedge [\tau]Z$ (expressing $[\![\downarrow]\!]\,\Phi$)[4]. These formulas generate different approximants.

$$
\begin{array}{llllll}
\nu Z^{0}.\,\Psi & = & \mathtt{tt} & \mu Z^{0}.\,\Psi & = & \mathtt{ff} \\
\nu Z^{1}.\,\Psi & = & \Phi \wedge [\tau]\mathtt{tt} = \Phi & \mu Z^{1}.\,\Psi & = & \Phi \wedge [\tau]\mathtt{ff} \\
\nu Z^{2}.\,\Psi & = & \Phi \wedge [\tau]\Phi & \mu Z^{2}.\,\Psi & = & \Phi \wedge [\tau](\Phi \wedge [\tau]\mathtt{ff}) \\
\vdots & & \vdots & \vdots & & \vdots
\end{array}
$$

[4] It is assumed that Z is not free in Φ.

$$\nu Z^i.\Psi = \Phi \wedge [\tau](\Phi \wedge [\tau](\Phi \wedge \ldots [\tau]\Phi \ldots))$$
$$\mu Z^i.\Psi = \Phi \wedge [\tau](\Phi \wedge [\tau](\Phi \wedge \ldots [\tau](\Phi \wedge [\tau]\mathtt{ff}) \ldots))$$
$$\vdots \qquad \vdots$$

The approximant $\mu Z^i.\Psi$ carries the extra demand that there cannot be a sequence of silent actions of length i. Hence, $[\![\downarrow]\!]\,\Phi$ requires all immediate τ behaviour to eventually peter out.

Exercises

1. Let P be the set of processes P(Ven). Using approximants, determine the sets $\| \Phi \|^{\mathsf{P}}_{\mathsf{V}}$ when Φ is each of the following.
   a. $\mu Z.\langle 2\mathrm{p}, 1\mathrm{p}\rangle \mathtt{tt} \vee [-]Z$
   b. $\mu Z.\langle \mathtt{little}\rangle \mathtt{tt} \vee [-]Z$
   c. $\mu Z.\langle \mathtt{little}\rangle \mathtt{tt} \vee [-]Z$
   d. $\nu Z.[2\mathrm{p}](\mu Y.\langle \mathtt{collect_b}\rangle \mathtt{tt} \vee [-]Y) \wedge [-]Z$
2. Let P be the set of processes P(Crossing). Using approximants, determine the sets $\| \Phi \|^{\mathsf{P}}_{\mathsf{V}}$ when Φ is each of the following.
   a. $\nu Z.([\overline{\mathtt{tcross}}]\mathtt{ff} \vee [\overline{\mathtt{ccross}}]\mathtt{ff}) \wedge [-]Z$
   b. $\mu Y.\langle -\rangle \mathtt{tt} \wedge [-\overline{\mathtt{ccross}}]Y$
3. If $|\mathsf{P}| = n$ prove that for any $E \in \mathsf{P}$
   a. $E \models_{\mathsf{V}} \nu Z.\Phi$ iff $E \models_{\mathsf{V}} \nu Z^n.\Phi$
   b. $E \models_{\mathsf{V}} \mu Z.\Phi$ iff $E \models_{\mathsf{V}} \mu Z^n.\Phi$
4. Work out the approximants $\sigma Z^i.\Phi$ for the following formulas for $i \leq 4$.
   a. $\mu Z.[-]Z$
   b. $\nu Z.[-]\langle \mathtt{tick}\rangle Z$
   c. $\mu Z.[-]Z \wedge [-][-]Z$
   d. $\nu Z.\langle \mathtt{tick}\rangle Z \wedge \langle \mathtt{tock}\rangle Z$
   e. $\mu Z.\Psi \vee (\Phi \wedge (\langle -\rangle \mathtt{tt} \wedge [-]Z))$
   f. $\nu Z.\Psi \vee (\Phi \wedge (\langle -\rangle \mathtt{tt} \wedge [-]Z))$
   g. $\mu Z.\nu Y.[a]Z \wedge [-a]Y$
5. Prove Proposition 2 part 2.

5.6 Embedded approximants

Fixed point sets can be calculated iteratively using approximants. The examples in the previous section involved a single fixed point. In this section, we examine the iterative technique in the presence of multiple fixed points, and comment on entanglement of approximants.

Example 1 Ven has the property $\nu Z. [\texttt{2p}, \texttt{1p}]\Psi \wedge [-]Z$, when Ψ is the formula $\mu Y. \langle - \rangle \texttt{tt} \wedge [-\{\texttt{collect}_\texttt{b}, \texttt{collect}_\texttt{l}\}]Y$; see example 1 of Section 5.1. Let P be the set {Ven, $\texttt{Ven}_\texttt{b}$, $\texttt{Ven}_\texttt{l}$, $\texttt{collect}_\texttt{b}$.Ven, $\texttt{collect}_\texttt{l}$.Ven}. First, the subset of P is calculated for the embedded fixed point Ψ, as follows. Its ith approximant is represented as μY^i (where we drop the index P).

$$\begin{aligned}
\mu Y^0 &= \emptyset \\
\mu Y^1 &= \| \langle - \rangle \texttt{tt} \wedge [-\{\texttt{collect}_\texttt{b}, \texttt{collect}_\texttt{l}\}]Y \|_{\mathsf{V}[\mu Y^0/Y]} \\
&= \{\texttt{collect}_\texttt{b}.\texttt{Ven}, \texttt{collect}_\texttt{l}.\texttt{Ven}\} \\
\mu Y^2 &= \| \langle - \rangle \texttt{tt} \wedge [-\{\texttt{collect}_\texttt{b}, \texttt{collect}_\texttt{l}\}]Y \|_{\mathsf{V}[\mu Y^1/Y]} \\
&= \{\texttt{Ven}_\texttt{b}, \texttt{Ven}_\texttt{l}, \texttt{collect}_\texttt{b}.\texttt{Ven}, \texttt{collect}_\texttt{l}.\texttt{Ven}\} \\
\mu Y^3 &= \| \langle - \rangle \texttt{tt} \wedge [-\{\texttt{collect}_\texttt{b}, \texttt{collect}_\texttt{l}\}]Y \|_{\mathsf{V}[\mu Y^2/Y]} \\
&= \mathsf{P}
\end{aligned}$$

Therefore, $\| \Psi \|_\mathsf{V}$ is P. Next, the outermost fixed point is evaluated, given that every process in P has the property Ψ. Its ith approximant is νZ^i.

$$\begin{aligned}
\nu Z^0 &= \mathsf{P} \\
\nu Z^1 &= \| [\texttt{2p}, \texttt{1p}]\Psi \wedge [-]Z \|_{\mathsf{V}[\nu Z^0/Z]} = \mathsf{P}
\end{aligned}$$

In this example, the embedded fixed point can be evaluated independently of the outermost fixed point.

Example 1 illustrates how the iterative technique works for formulas with multiple fixed points that are independent of each other. The formula of example 1 has the form $\nu Z. \Phi(Z, \mu Y. \Psi(Y))$ where the notation makes explicit what variables can be free in subformulas. Z does not occur free within the subformula $\mu Y. \Psi(Y)$, but may occur within the subformula $\Phi(Z, \mu Y. \Psi(Y))$. Consequently, when evaluating the outermost fixed point, we have that:

$$\begin{aligned}
\nu Z^0 &= \mathsf{P} \\
\nu Z^1 &= \| \Phi(Z, \mu Y. \Psi(Y)) \|_{\mathsf{V}[\nu Z^0/Z]} \\
&\vdots \\
\nu Z^{i+1} &= \| \Phi(Z, \mu Y. \Psi(Y)) \|_{\mathsf{V}[\nu Z^i/Z]} \\
&\vdots
\end{aligned}$$

Throughout these approximants, the subset of processes with the property $\mu Y. \Psi(Y)$ is invariant. This is because the subformula does not contain Z free. Therefore, $\| \mu Y. \Psi(Y) \|_{\mathsf{V}[\nu Z^\alpha/Z]}$ is the same set as $\| \mu Y. \Psi(Y) \|_{\mathsf{V}[\nu Z^\beta/Z]}$ for any ordinals α and β.

The subset of formulas of μM, written μMI, which has the property that all its fixed points are independent of each other, is characterised as follows.

$$\Phi \in \mu\text{MI} \quad \text{iff} \quad \text{if } \sigma_1 Y.\, \Psi_1 \in \text{Sub}(\Phi) \text{ and } \sigma_2 Z.\, \Psi_2 \in \text{Sub}(\Phi) \text{ and } Y \neq Z, \\ \text{then } Y \text{ is not free in } \Psi_2 \text{ and } Z \text{ is not free in } \Psi_1.$$

CTL formulas, when understood as fixed point formulas, belong to this subclass of μM formulas.

Proposition 1 *CTL formulas belong to μMI.*

Proof. This follows from the fixed point definitions of until formulas (and their negations) presented in Section 5.2. For example, A(Φ U Ψ) is defined as $\mu Z.\, \Psi \vee (\Phi \wedge (\langle - \rangle \mathtt{tt} \wedge [-]Z))$ where Z does not occur in Φ or Ψ. Therefore, any further fixed points introduced into these subformulas do not contain Z. □

Example 2 Consider the simple processes in Figure 5.1. Let P be the set $\{\mathtt{D}, \mathtt{D}', \mathtt{D}''\}$ and let Ψ be the formula

$$\Psi \quad = \quad \mu Y.\, \nu Z.\, [a]((\langle b \rangle \mathtt{tt} \vee Y) \wedge Z).$$

This formula does not belong to μMI because its innermost fixed point subformula $\nu Z \ldots$ contains Y free. The calculation of the outer fixed point depends on calculating the inner fixed point at each index. We use νZ^{ji} to represent the ith approximant of the subformula prefaced with νZ when any free occurrence of Y

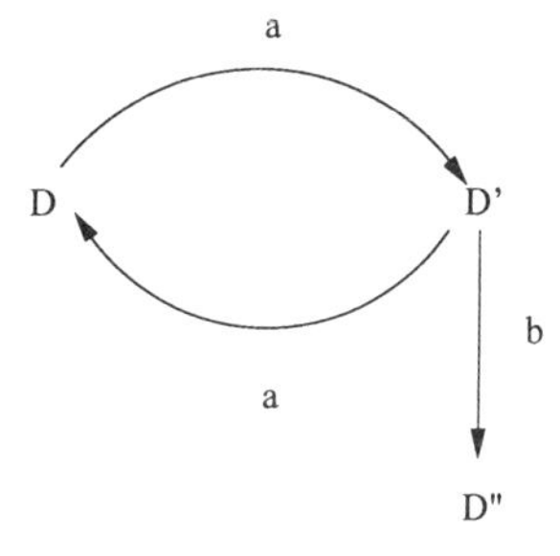

FIGURE 5.1. A simple process

is understood as the approximant μY^j. Again, we drop the index P.

$$
\begin{aligned}
\mu Y^0 &= \emptyset \\
\mu Y^1 &= \| \nu Z.\, [a]((\langle b\rangle \mathtt{tt} \vee Y) \wedge Z) \|_{\mathsf{V}[\mu Y^0/Y]} \\
& \quad \nu Z^{00} = \mathsf{P} \\
& \quad \nu Z^{01} = \| [a]((\langle b\rangle \mathtt{tt} \vee Y) \wedge Z) \|_{(\mathsf{V}[\mu Y^0/Y])[\nu Z^{00}/Z]} \\
& \quad\quad\quad = \{\mathtt{D}'', \mathtt{D}\} \\
& \quad \nu Z^{02} = \| [a]((\langle b\rangle \mathtt{tt} \vee Y) \wedge Z) \|_{(\mathsf{V}[\mu Y^0/Y])[\nu Z^{01}/Z]} \\
& \quad\quad\quad = \{\mathtt{D}''\} \\
& \quad \nu Z^{03} = \| [a]((\langle b\rangle \mathtt{tt} \vee Y) \wedge Z) \|_{(\mathsf{V}[\mu Y^0/Y])[\nu Z^{02}/Z]} \\
& \quad\quad\quad = \{\mathtt{D}''\} \\
\text{So } \mu Y^1 &= \{\mathtt{D}''\} \\
\mu Y^2 &= \| \nu Z.\, [a]((\langle b\rangle \mathtt{tt} \vee Y) \wedge Z) \|_{\mathsf{V}[\mu Y^1/Y]} \\
& \quad \nu Z^{10} = \mathsf{P} \\
& \quad \nu Z^{11} = \| [a]((\langle b\rangle \mathtt{tt} \vee Y) \wedge Z) \|_{(\mathsf{V}[\mu Y^1/Y])[\nu Z^{10}/Z]} \\
& \quad\quad\quad = \{\mathtt{D}'', \mathtt{D}\} \\
& \quad \nu Z^{12} = \| [a]((\langle b\rangle \mathtt{tt} \vee Y) \wedge Z) \|_{(\mathsf{V}[\mu Y^1/Y])[\nu Z^{11}/Z]} \\
& \quad\quad\quad = \{\mathtt{D}''\} \\
& \quad \nu Z^{13} = \| [a]((\langle b\rangle \mathtt{tt} \vee Y) \wedge Z) \|_{(\mathsf{V}[\mu Y^1/Y])[\nu Z^{12}/Z]} \\
& \quad\quad\quad = \{\mathtt{D}''\} \\
\text{So } \mu Y^2 &= \{\mathtt{D}''\}
\end{aligned}
$$

The innermost fixed point is evaluated with respect to more than one outermost approximant.

Example 2 illustrates dependence of fixed points. The formula has the form $\mu Y.\, \Phi(Y, \nu Z.\, \Psi(Y, Z))$, where Y is free in the innermost fixed point. The interpretation of the subformula $\nu Z.\, \Psi(Y, Z)$ may vary according to the interpretation of Y.

A simple measure of how much work has to be done when calculating the subset of P with a fixed point property is the number of approximants that must be calculated. For a simple formula $\sigma Z.\, \Phi$, when Φ does not contain fixed points and when $|\mathsf{P}| = n$, the maximum number of approximants is n. If $\Phi \in \mu\mathrm{MI}$ and contains k fixed points, then the maximum calculation needed is $k \times n$. In the case of the formula of example 2 with two fixed points, it appears that n^2 is the upper bound. For each outermost approximant, we may have to calculate n inner approximants. There are at most n outer approximants, and therefore the upper bound is n^2.

The initial approximants for fixed points are $\nu Z^0 = \mathsf{P}$ and $\mu Z^0 = \emptyset$. Initial approximations that are closer to the required fixed point will, in general, mean that less work has to be done in the calculation. For instance, if we know that E has the property $\mathsf{P} \supseteq \mathsf{E} \supseteq \| \nu Z.\, \Phi \|_{\mathsf{V}}$, then we can set νZ^0 to E. This observation can be used when evaluating the embedded fixed point formula $\nu Z.\, \Phi(Z, \nu Y.\, \Psi(Z, Y))$.

$$
\begin{array}{lll}
\nu Z^0 & = & \mathsf{P} \\
\nu Z^1 & = & \| \Phi(Z, \nu Y.\, \Psi(Z, Y)) \|_{\mathsf{V}[\nu Z^0/Z]} \\
 & & \quad \nu Y^{00} = \mathsf{P} \\
 & & \quad \nu Y^{01} = \| \Psi(Z, Y) \|_{(\mathsf{V}[\nu Z^0/Z])[\nu Y^{00}/Y]} \\
 & & \quad \vdots \\
\nu Z^2 & = & \| \Phi(Z, \nu Y.\, \Psi(Z, Y)) \|_{\mathsf{V}[\nu Z^1/Z]}
\end{array}
$$

To evaluate νZ^2 one needs to calculate $\| \nu Y.\, \Psi(Z, Y) \|_{\mathsf{V}[\nu Z^1/Z]}$. However, by monotonicity we know that

$$\| \nu Y.\, \Psi(Z, Y) \|_{\mathsf{V}[\nu Z^1/Z]} \subseteq \| \nu Y.\, \Psi(Z, Y) \|_{\mathsf{V}[\nu Z^0/Z]} \subseteq \mathsf{P}.$$

Therefore, we can use $\| \nu Y.\, \Psi(Z, Y) \|_{\mathsf{V}[\nu Z^0/Z]}$ as the initial approximant νY^{10}, and so on as follows.

$$
\begin{array}{lll}
\nu Z^2 & = & \| \Phi(Z, \nu Y.\, \Psi(Z, Y)) \|_{\mathsf{V}[\nu Z^1/Z]} \\
 & & \quad \nu Y^{10} = \| \nu Y.\, \Psi(Z, Y) \|_{\mathsf{V}[\nu Z^0/Z]} \\
 & & \quad \nu Y^{11} = \| \Psi(Z, Y) \|_{(\mathsf{V}[\nu Z^1/Z])[\nu Y^{10}/Y]} \\
 & & \quad \vdots \\
\nu Z^{i+1} & = & \| \Phi(Z, \nu Y.\, \Psi(Z, Y)) \|_{\mathsf{V}[\nu Z^i/Z]} \\
 & & \quad \nu Y^{i0} = \| \nu Y.\, \Psi(Z, Y) \|_{\mathsf{V}[\nu Z^i/Z]} \\
 & & \quad \nu Y^{i1} = \| \Psi(Z, Y) \|_{(\mathsf{V}[\nu Z^i/Z])[\nu Y^{i0}/Y]} \\
 & & \quad \vdots \\
\vdots & &
\end{array}
$$

Consequently, instead of requiring at most n^2 approximants, the upper bound is $2 \times n$, for in total the maximum number of approximants for the inner fixed point is n. This technique can be extended to long sequences of embedded maximal fixed points

$$\nu Z_1.\, \Phi_1(Z_1, \nu Z_2.\, \Phi_2(Z_1, Z_2, \ldots, \nu Z_k.\, \Phi_k(Z_1, \ldots, Z_k)\ldots))$$

when at most $k \times n$ approximants need to be calculated.

The situation is dual for least fixed points. If $\emptyset \subseteq \mathsf{E} \subseteq \| \mu Z.\, \Phi \|_{\mathsf{V}}$, then we can set the initial set μZ^0 to be E. In the case of the formula $\mu Z.\, \Phi(Z, \mu Y.\, \Psi(Z, Y))$, by monotonicity

$$\emptyset \subseteq \| \mu Y.\, \Psi(Z, Y) \|_{\mathsf{V}[\mu Z^i/Z]} \subseteq \| \mu Y.\, \Psi(Z, Y) \|_{\mathsf{V}[\mu Z^{i+1}/Z]},$$

so the set $\| \mu Y.\, \Psi(Z, Y) \|_{V[\mu Z^i/Z]}$ is a better initial approximation than $\emptyset$ for μY^{i0}. Again, we need only calculate $2 \times n$ approximants instead of n^2. This can be extended to multiple occurrences of embedded least fixed points.

$$\mu Z_1.\, \Phi_1(Z_1, \mu Z_2.\, \Phi_2(Z_1, Z_2, \ldots \mu Z_k.\, \Phi_k(Z_1, \ldots, Z_k) \ldots))$$

More complex are formulas, as in example 2, that contain multiple occurrences of dependent but different fixed points. In the case of a formula $\nu Z.\, \Phi(Z, \mu Y.\, \Psi(Z, Y))$, we cannot use $\| \mu Y.\, \Psi(Z, Y) \|_{V[\nu Z^0/Z]}$ as an initial approximant μY^{10} because the ordering is the wrong way around to be of help, since $\nu Z^0 \supseteq \nu Z^1$. Least fixed points are approximated from below and not from above. Similar comments apply to the calculation of $\mu Z.\, \Phi(Z, \nu Y.\, \Psi(Z, Y))$. In both these cases, a best worst case estimate of the number of approximants is n^2. However, when we consider three fixed points

$$\mu Y.\, \Phi_1(Y, \nu Z.\, \Phi_2(Y, Z, \mu X.\, \Phi_3(Y, Z, X))),$$

the maximum number of calculations needed is less than n^3. As the reader can verify montonicity can be used on the innermost fixed point because $\mu X^{ijk} \subseteq \mu X^{i'jk}$ when $i \leq i'$. Roughly speaking, the number of calculations required for k dependent alternating fixed points is $n^{\frac{k+1}{2}}$, as was proved by Long et al [38].

The general issue of how many approximants are needed to calculate an embedded fixed point remains an open problem. The observations here suggest that the more alternating dependent fixed points there are, the more calculations are needed, and that the number of calculations is exponential in this alternation. Indeed, we may wonder if somehow the expressive power of μM grows as more dependent alternating fixed points are permitted. The answer is "yes," as was proved by Bradfield [9].

A more generous subset of μM than μMI is the alternation-free fragment, μMA defined as follows.

$\Phi \in \mu\text{MA}$ iff if $\mu Y.\, \Psi_1 \in \text{Sub}(\Phi)$ and $\nu Z.\, \Psi_2 \in \text{Sub}(\Phi)$, then Y is not free in Ψ_2 and Z is not free in Ψ_1.

This subset permits dependence of fixed points, provided they are of the same kind. For instance, a formula of the following form is permitted.

$$\nu Z_1.\, \Phi_1(Z_1, \nu Z_2.\, \Phi_2(Z_1, Z_2, \mu Y_1.\, \Psi_1(Y_1, \mu Y_2.\, \Psi_2(Y_1, Y_2))))$$

An example of a formula that does not belong to μMA is Ψ from example 2, since Y is free in the subformula $\nu Z \ldots$. From the analysis above, any formula of μMA with k fixed point operators requires the calculation of at most $k \times n$ approximants.

Exercises

1. Show precisely that, if $|\mathsf{P}| = n$, then for both formulas

$$\nu Z_1.\, \Phi_1(Z_1, \nu Z_2.\, \Phi_2(Z_1, Z_2, \ldots, \nu Z_k.\, \Phi_k(Z_1, \ldots, Z_k) \ldots))$$
$$\mu Z_1.\, \Phi_1(Z_1, \mu Z_2.\, \Phi_2(Z_1, Z_2, \ldots \mu Z_k.\, \Phi_k(Z_1, \ldots, Z_k) \ldots))$$

at most $k \times n$ approximants need to be calculated.

2. For the following formulas, work out their approximants and embedded approximants up to stage 4.

   a. $\mu Y.\, [b]\mathtt{ff} \wedge [a](\mu Z.\, [a]\mathtt{ff} \wedge [b]Y \wedge [-b]Z) \wedge [-a]Y$

   b. $\nu Y.\, [b]\mathtt{ff} \wedge [a](\mu Z.\, [a]\mathtt{ff} \wedge [b]Y \wedge [-b]Z) \wedge [-a]Y$

   c. $\mu X.\, [a](\mu Y.\, [a](\nu Z.\, [a]\mathtt{ff} \wedge [-]Z) \wedge \langle - \rangle \mathtt{tt} \wedge [-a]Y) \wedge \langle - \rangle \mathtt{tt} \wedge [-a]X$

   d. $\mu X.\, \nu Y.\, [a]X \wedge [-a]Y$

   e. $\nu Z.\, [a](\mu Y.\, \langle - \rangle \mathtt{tt} \wedge [-b]Y) \wedge [-]Z$

   f. $\nu Z.\, (\Psi^c \vee (\Psi \wedge \nu Y.\, \Phi \wedge [-]Y)) \wedge [-]Z$

   g. $\nu Z.\, (\mu X.\, [b](\nu Y.\, [c](\nu Y_1.\, X \wedge [-a]Y_1) \wedge [-a]Y) \wedge [-]Z)$

3. Give an exact upper bound on the number of approximants that need to be calculated in the case of the following formula with respect to $|\mathsf{P}| = n$.

   $$\mu Y.\, \Phi_1(Y, \nu Z.\, \Phi_2(Y, Z, \mu X.\, \Phi_3(Y, Z, X)))$$

   What upper bound is there for a formula with k dependent alternating fixed points?

4. Prove Proposition 1 in full. Show using approximants that there is a linear time algorithm for checking whether $E \models \Phi$, when E is finite state and Φ is a CTL formula.

5. Show that, for any $\Phi \in \mu$MA with k fixed point operators, at most $k \times n$ approximants need to be calculated to work out which subset of P has property Φ when $|\mathsf{P}| = n$.

6. The following is a technical definition of the full alternation hierarchy from Niwinski and Bradfield [46, 9].

   a. If Φ contains no fixed point operators, then $\Phi \in \Sigma_0$ and $\Phi \in \Pi_0$

   b. If $\Phi \in \Sigma_n \cup \Pi_n$, then $\Phi \in \Sigma_{n+1}$ and $\Phi \in \Pi_{n+1}$

   c. If $\Phi, \Psi \in \Sigma_n(\Pi_n)$, then $[K]\Phi$, $\langle K \rangle \Phi$, $\Phi \wedge \Psi$, $\Phi \vee \Psi \in \Sigma_n(\Pi_n)$

   d. If $\Phi \in \Sigma_n$, then $\mu Z.\, \Phi \in \Sigma_n$

   e. If $\Phi \in \Pi_n$, then $\nu Z.\, \Phi \in \Pi_n$.

   f. If $\Phi, \Psi \in \Sigma_n(\Pi_n)$, then $\Phi\{\Psi/Z\} \in \Sigma_n(\Pi_n)$

   Let AD_n be $\{\Phi \,:\, \Phi \in \Sigma_{n+1} \cap \Pi_{n+1}\}$.

   a. Prove that $\mathrm{AD}_1 = \mu$MA.

   b. Give an example of a property that is definable in AD_{n+1}, but not definable in AD_n. (See Bradfield [9].)

   c. Provide an upper bound for the number of approximants that need to be calculated for any $\Phi \in \mathrm{AD}_m$ containing k fixed points, when $|\mathsf{P}| = n$.

5.7 Expressing properties

Modal mu-calculus is a very powerful temporal logic that permits expression of a very rich class of properties. In this section, we examine how to express a range of liveness and safety properties.

A safety property, as described in the previous chapter, has the form "nothing bad ever happens." Safety can either be ascribed to "states" (or "processes"), that bad states can never be reached, or to "actions," that bad actions never happen. In the former case, if the formula Φ^c captures the bad states, then the CTL formula $\mathrm{AG}\Phi$ or its μM equivalent $\nu Z.\, \Phi \wedge [-]Z$ expresses safety. As mentioned before, the safety property for the crossing of Figure 1.10 is that it is never possible to reach a state such that a train and a car are both able to cross, $\mathrm{AG}([\overline{\mathtt{tcross}}]\mathtt{ff} \vee [\overline{\mathtt{ccross}}]\mathtt{ff})$.

It is useful to allow the full freedom of μM notation by allowing open formulas with free variables and appropriate valuations that capture their intended meaning. In the case of a safety property, let E be the family of bad states. The formula $\nu Z.\, Q \wedge [-]Z$ expresses safety relative to the valuation V that assigns P − E to Q. The idea is that the free variable Q has a definite intended meaning captured by the particular valuation V.

Example 1 The slot machine of Figure 1.15 never has a negative amount of money. A simple way of expressing this feature is the open formula $\nu Z.Q \wedge [-]Z$ relative to the valuation V that assigns to the free variable Q the set $\mathsf{P} - \{\mathtt{SM}_j : j < 0\}$.

This device can be used to express properties succinctly. If the particular valuation V is bisimulation closed with respect to the free variables of Φ, we say that the property expressed by Φ relative to V is *extensional*. In this case $\| \Phi \|_{\mathsf{V}}$ is also bisimulation closed by Proposition 3 of Section 5.4. The formula of example 1 is extensional when P is the set of processes of the transition graph for $\mathtt{SM}_n$ for $n \geq 0$. If Φ relative to V is not extensional, then it is intensional. Examples of intensional properties of a process include "has two `tick`-transitions" and "consists of three parallel components."

Safety can also be ascribed to actions, "no bad action belonging to K ever happens." It is expressed by the μM formula $\nu Z.\, [K]\mathtt{ff} \wedge [-]Z$ (or the equivalent CTL formula $\mathrm{AG}[K]\mathtt{ff}$). Really, there is no distinction between safety in terms of bad states and safety in terms of bad actions. In the action case, safety is equivalent to "the bad state $\langle K\rangle\mathtt{tt}$ is never reached."

A liveness property has the form "something good eventually happens." Again, the good feature can be ascribed either to states or to actions. If Φ is true at the good states, then the CTL formula $\mathrm{AF}\Phi$ or its μM equivalent $\mu Z.\, \Phi \vee (\langle -\rangle\mathtt{tt} \wedge [-]Z)$ expresses liveness.

In contrast, that eventually some action in K happens means that all possible runs contain a transition whose action belongs to K. This liveness property is expressed by the following formula.

$$\mu Z.\,\langle - \rangle \mathtt{tt} \wedge [-K]Z$$

To satisfy it, a process must not be able to perform actions from the set $\mathsf{A} - K$ forever. Consider the syntactic approximations (as described in Section 5.5) this formula generates. We abbreviate the ith approximant to μZ^i.

$$\begin{array}{lcl}
\mu Z^0 & = & \mathtt{ff} \\
\mu Z^1 & = & \langle - \rangle \mathtt{tt} \wedge [-K]\mu Z^0 \;=\; \langle - \rangle \mathtt{tt} \wedge [-K]\mathtt{ff} \\
\mu Z^2 & = & \langle - \rangle \mathtt{tt} \wedge [-K]\mu Z^1 \;=\; \langle - \rangle \mathtt{tt} \wedge [-K](\langle - \rangle \mathtt{tt} \wedge [-K]\mathtt{ff}) \\
\vdots & & \vdots \\
\mu Z^{i+1} & = & \langle - \rangle \mathtt{tt} \wedge [-K]\mu Z^i \\
 & = & \langle - \rangle \mathtt{tt} \wedge [-K](\langle - \rangle \mathtt{tt} \wedge \ldots [-K](\langle - \rangle \mathtt{tt} \wedge [-K]\mathtt{ff}) \ldots) \\
\vdots & & \vdots
\end{array}$$

The approximant μZ^1 expresses that a K action must happen next. μZ^2 states that a K action must happen within two transitions, that is, for any sequence of transitions $\xrightarrow{ab}$ either $a \in K$ or $b \in K$. Moreover, the formula requires that, if there is a run with just one transition, then it is a K transition. Generalising, μZ^i states that any sequence of transitions of length i contains a K action (and any run whose transition length is less than i contains a K transition). Therefore, $\mu Z^{\omega+1} = \bigvee \{\mu Z^i \;:\; i \geq 0\}$ guarantees the liveness property that in every run there is a K action.

Liveness with respect to actions does not appear to be expressible in terms of liveness with respect to state, that is, as a formula of the form $\mu Z.\,\Phi \vee (\langle - \rangle \mathtt{tt} \wedge [-]Z)$ where Φ does not contain fixed points (or Z). For instance, neither of the following formulas captures liveness.

$$\begin{array}{lcl}
\Phi_1 & = & \mu Z.\,\langle K \rangle \mathtt{tt} \vee (\langle - \rangle \mathtt{tt} \wedge [-]Z) \\
\Phi_2 & = & \mu Z.\,(\langle - \rangle \mathtt{tt} \wedge [-K]\mathtt{ff}) \vee (\langle - \rangle \mathtt{tt} \wedge [-]Z)
\end{array}$$

Φ_1 is too weak, since it merely states that eventually some action in K is possible without any guarantee that it happens. In contrast, Φ_2 is too strong, since it states that eventually only K actions are possible (and therefore must happen).

Example 2 Consider the following definitions of processes.

$$\begin{array}{lcl}
\mathtt{A}_0 & \stackrel{\text{def}}{=} & a.\sum\{\mathtt{A}_i \;:\; i \geq 0\} \\
\mathtt{A}_{i+1} & \stackrel{\text{def}}{=} & b.\mathtt{A}_i \quad i \geq 0 \\
\mathtt{B}_0 & \stackrel{\text{def}}{=} & a.\sum\{\mathtt{B}_i \;:\; i \geq 0\} + b.\sum\{\mathtt{B}_i \;:\; i \geq 0\} \\
\mathtt{B}_{i+1} & \stackrel{\text{def}}{=} & b.\mathtt{B}_i \quad i \geq 0
\end{array}$$

For any $i \geq 0$, process $\mathtt{A}_0$ can perform the cycle $\mathtt{A}_0 \xrightarrow{a\,b^i} \mathtt{A}_0$, whereas $\mathtt{B}_0$ can perform the cycles $\mathtt{B}_0 \xrightarrow{a\,b^i} \mathtt{B}_0$ and $\mathtt{B}_0 \xrightarrow{b\,b^i} \mathtt{B}_0$. When $j \geq 0$, $\mathtt{A}_j \mid \mathtt{A}_j$ has the property "eventually a happens," which is not shared by $\mathtt{B}_j \mid \mathtt{B}_j$. In every run of $\mathtt{A}_j \mid \mathtt{A}_j$, the action a occurs. There are runs from $\mathtt{B}_j \mid \mathtt{B}_j$ consisting only of b actions. Therefore, $\mathtt{A}_j \mid \mathtt{A}_j \models \mu Z.\, \langle - \rangle \mathtt{tt} \wedge [-a]Z$ and $\mathtt{B}_j \mid \mathtt{B}_j \not\models \mu Z.\, \langle - \rangle \mathtt{tt} \wedge [-a]Z$. On the other hand, both these processes have the property Φ_1, above, when K is the singleton set $\{a\}$. Process $\mathtt{B}_j \mid \mathtt{B}_j$ eventually reaches $\mathtt{B}_0 \mid \mathtt{B}_k$ or $\mathtt{B}_k \mid \mathtt{B}_0$, and both satisfy $\langle a \rangle \mathtt{tt}$. Moreover, both fail to have the property Φ_2, above, when K is the set $\{a\}$. For instance, in the case of $\mathtt{A}_j \mid \mathtt{A}_j$, there is no guarantee that in every run $\mathtt{A}_0 \mid \mathtt{A}_0$ is reached because this is the only process of the form $\mathtt{A}_i \mid \mathtt{A}_j$ satisfying $\langle - \rangle \mathtt{tt} \wedge [-a]\mathtt{ff}$.

Liveness and safety may relate to subsets of runs. For instance, they may be triggered by particular actions or states. A simple case is "if action a ever happens, then eventually b happens," any run with an a action has a later b action. This is expressed by the following formula.

$$\nu Z.\, [a](\mu Y.\, \langle - \rangle \mathtt{tt} \wedge [-b]Y) \wedge [-]Z$$

An example of a conditional safety property is "whenever Ψ holds, Φ^c will never become true," which is expressed as follows.

$$\nu Z.\, (\Psi^c \vee (\Psi \wedge \nu Y.\, \Phi \wedge [-]Y)) \wedge [-]Z$$

In both these examples, the formulas belong to μMI (as defined in the previous section).

More involved is the expression of liveness properties under fairness. An example is "in any run, if b and c happen infinitely often, then so does a," expressed as follows.

$$\nu Z.\, (\mu X.\, [b](\nu Y.\, [c](\nu Y_1.\, X \wedge [-a]Y_1) \wedge [-a]Y) \wedge [-]Z)$$

There is an essential fixed point dependence, as described in the previous section, because X occurs free within the fixed point subformula prefaced with νY.

Example 3 The desirable liveness property for the crossing "whenever a car approaches the crossing, eventually it crosses" is captured by the following formula.

$$\nu Z.\, [\mathtt{car}](\mu Y.\, \langle - \rangle \mathtt{tt} \wedge [-\overline{\mathtt{ccross}}]Y) \wedge [-]Z$$

However, this only holds if we assume that the signal is fair. Let Q and R be variables, and V a valuation, such that Q is true when the crossing is in any state where Rail has the form $\mathtt{green}.\overline{\mathtt{tcross}}.\overline{\mathtt{red}}.\mathtt{Rail}$ (the states E_2, E_3, E_6, and E_{10} of figure 1.12) and R holds when it is in any state where Road has the form $\mathtt{up}.\overline{\mathtt{ccross}}.\overline{\mathtt{down}}.\mathtt{Road}$ (the states E_1, E_3, E_7 and E_{11}). The liveness property now becomes "for any run, if Q^c is true infinitely often and R^c is also true infinitely often, then whenever a car approaches the crossing, eventually it crosses," which is expressed by the open formula relative to V,

$$\nu Y.\,[\mathtt{car}](\mu X.\,\nu Y_1.\,(Q \vee [-\overline{\mathtt{ccross}}](\Psi \wedge [-\overline{\mathtt{ccross}}]Y_1))) \wedge [-]Y,$$

where Ψ is $\nu Y_2.\,(R \vee X) \wedge [-\{\overline{\mathtt{ccross}}\}]Y_2$. The property expressed here is extensional.

Another class of properties is until properties, as in CTL. The formula $\mathrm{A}(\Phi\ \mathrm{U}\ \Psi)$ expresses "for any run, Φ holds until Ψ becomes true," which in μM is the formula $\mu Z.\,\Psi \vee (\Phi \wedge \langle - \rangle \mathtt{tt} \wedge [-]Z)$, where Z does not occur in Φ or Ψ. This property does require that Ψ eventually become true. This commitment can be removed by changing fixed points. The property "in every run, Φ holds unless Ψ becomes true" does not imply that Ψ does eventually hold, and therefore is expressed as $\nu Y.\,\Psi \vee (\Phi \wedge [-]Y)$. Until properties may concern actions instead of states. For example, "in any run, K actions happen until a J action happens" is expressed as $\mu Y.\,[-(K \cup J)]\mathtt{ff} \wedge \langle - \rangle \mathtt{tt} \wedge [-J]Y$, with the implication that eventually a J action occurs. This implication can be removed, as in the property "in any run, K actions happen unless a J action occurs," by changing fixed points $\nu Y.\,[-(K \cup J)]\mathtt{ff} \wedge [-J]Y$. The reader is invited to formulate weaker versions of these properties with respect to some runs.

Cyclic properties can also be described in μM. A simple example is that tock recurs at each even point: if $E_0 \xrightarrow{a_1} E_1 \xrightarrow{a_2} \ldots$ is a finite or infinite length run, then each a_{2i} is tock. The formula $\nu Z.\,[-]([-\mathtt{tock}]\mathtt{ff} \wedge [-]Z)$ expresses this property, and $\mathtt{Cl}_1$ satisfies it. The clock $\mathtt{Cl}_1$ in fact has an even more regular cyclic property, that every run is a repeating cycle of tick and tock actions, $\nu Z.\,[-\mathtt{tick}]\mathtt{ff} \wedge [\mathtt{tick}]([-\mathtt{tock}]\mathtt{ff} \wedge [-]Z)$[5]. These properties can also be weakened to some family of runs. Cyclic properties that allow other actions to intervene within a cycle can also be expressed.

Example 4 Recall the scheduler from Section 1.4 that schedules a sequence of tasks, and must ensure that a task cannot be restarted until its previous operation has finished. Suppose that initiation of one of the tasks is given by the action a and its termination by b. The scheduler therefore has to guarantee the cyclic behaviour $a\ b$ when other actions may occur before and after each occurrence of a and each occurrence of b. This property can be defined inductively as follows.

$$\mathrm{Cycle}(ab) \stackrel{\mathrm{def}}{=} [b]\mathtt{ff} \wedge [a]\mathrm{Cycle}(ba) \wedge [-a]\mathrm{Cycle}(ab)$$

$$\mathrm{Cycle}(ba) \stackrel{\mathrm{def}}{=} [a]\mathtt{ff} \wedge [b]\mathrm{Cycle}(ab) \wedge [-b]\mathrm{Cycle}(ba)$$

Here we have left open the possibility that runs have finite length. Appropriate occurrences of $\langle - \rangle \mathtt{tt}$ within the definition preclude it. An important issue is whether these recursive definitions are to be interpreted with least or greatest fixed points,

[5]This formula leaves open the possibility that a run has finite length. To preclude it, we add $\langle - \rangle \mathtt{tt}$ at the outer and inner levels.

or a mixture of the two. This depends upon whether intervening actions are allowed to continue forever without the next a or b happening. If we prohibit this, the cyclic property is expressed using least fixed points.

$$\mu Y. [b]\mathtt{ff} \wedge [a](\mu Z. [a]\mathtt{ff} \wedge [b]Y \wedge [-b]Z) \wedge [-a]Y$$

If we permit other actions to intervene forever, but insist that whenever a happens b happens later, a mixture of fixed points is required.

$$\nu Y. [b]\mathtt{ff} \wedge [a](\mu Z. [a]\mathtt{ff} \wedge [b]Y \wedge [-b]Z) \wedge [-a]Y$$

The length of the cycle can be extended. As an exercise the reader is invited to define the formula for Cycle($abcd$).

Another class of properties involves counting. An instance is that, in each run there are exactly two a actions, given as follows.

$$\mu X. [a](\mu Y. [a](\nu Z. [a]\mathtt{ff} \wedge [-]Z) \wedge \langle - \rangle \mathtt{tt} \wedge [-a]Y) \wedge \langle - \rangle \mathtt{tt} \wedge [-a]X$$

Another property is that, in every run, there are at least two a actions. Even more general is the property "in each run, a can only happen finitely often," which is expressed by $\mu X. \nu Y. [a]X \wedge [-a]Y$. However, there are also many counting properties that are not expressible in the logic. A notable case is the following property of a buffer, "the number of `out` actions never exceeds the number of `in` actions."

Exercises

1. Define in μM the following properties
   a. Eventually either `tick` happens or Φ becomes true
   b. In some run Φ is always true
   c. `tick` happens until Φ
   d. `tick` happens until `tock`
   e. `tick` happens unless `tock` happens
2. Prove that liveness with respect to actions cannot be expressed in terms of liveness with respect to state, that is, as a formula of the form $\mu Z. \Phi \vee (\langle - \rangle \mathtt{tt} \wedge [-]Z)$, where Φ is a modal formula that does not contain fixed points, or the free variable Z.
3. Define in μM the following properties.
   a. In any run, a and b happen finitely often
   b. If a, b and c happen infinitely often, then Φ is true infinitely often
   c. In any run, Φ is true twice and Ψ is true twice
4. Define fixed point formulas for the properties Cycle($abcd$), which depend on assumptions about intervening actions.
5. Prove that "the number of `out` actions never exceeds the number of `in` actions" is not expressible in μM (see Sistla et al. [52] for a proof technique).

6

Verifying Temporal Properties

A very rich temporal logic, modal mu-calculus, has been described. Formulas of the logic can express liveness, safety, cyclic and other properties of processes. The next step is to provide techniques for verification, for showing when processes have, or fail to have, these features. In this chapter, we show that game theoretic ideas provide a general framework for verification.

6.1 Techniques for verification

To show that a process has, or fails to have, a modal property we can appeal to the inductive definition of satisfaction between a process and a formula. A simple approach is goal directed. We start with the goal, $E \models \Phi$?, that is, "does E satisfy Φ?" and then continue reducing goals to subgoals until we reach either "obviously true" or "obviously false" subgoals. The reduction of goals to subgoals can proceed via rules that depend on the main connective of the formula in the goal. For example, the goal

$$\text{Goal}: \quad E \models [K]\Phi \ ?$$

reduces to the subgoals

$$\text{Subgoal}: \quad F \models \Phi \;?$$

for each F such that $E \xrightarrow{a} F$ and $a \in K$. Similarly, the next goal

$$\text{Goal}: \quad E \models \Phi_1 \vee \Phi_2 \;?$$

reduces to

$$\text{Subgoal}: \quad E \models \Phi_1 \;?$$

or to

$$\text{Subgoal}: \quad E \models \Phi_2 \;?$$

This technique has the merit that the formula Ψ in any subgoal is a proper subformula of the goal formula Φ, that is $\Psi \in \text{Sub}(\Phi)$. The method thereby supports the general principle that a proof of a property "follows from" subproofs of subproperties.

Checking whether a process satisfies a μM formula is not as straightforward. The problem is what to do with fixed point formulas, for instance with the following goal.

$$\text{Goal}: \quad E \models_{\mathsf{V}} \nu Z.\,\Phi \;?$$

According to the semantic clause for νZ in Section 5.1 the goal reduces to the subgoals

$$\text{Subgoal}: \quad F \models_{\mathsf{V}[\mathsf{E}/Z]} \Phi \;?$$

for each $F \in \mathsf{E}$ when E is a subset of $\mathsf{P}(E)$. This requires us first to identify the set $\mathsf{P}(E)$, and then to choose an appropriate subset E.

We can avoid calculating $\mathsf{P}(E)$ and choosing subsets of it by appealing to approximants, as presented in Sections 5.5 and 5.6. The goal now reduces to the subgoals

$$\text{Subgoal}: \quad E \models_{\mathsf{V}} \nu Z^{\alpha}.\,\Phi \;?$$

for each ordinal α. At this point we may use induction over ordinals. However, we need to be careful with limit ordinals. If the formula contains embedded fixed points, we may need to use simultaneous induction over ordinals. However, if E is a finite state process, then the goal reduces to the single subgoal

$$\text{Subgoal}: \quad E \models_{\mathsf{V}} \nu Z^{n}.\,\Phi \;?$$

where n is the number of processes in the transition graph for E (or an overestimate of its size). This allows one to reduce verification of a μM property to an M property.

However, it has the disadvantage that a proof of a property no longer follows from subproofs of subproperties.

Discovering a fixed point set in general is not easy, and is therefore prone to error. We therefore prefer simpler, and consequently safer, methods for checking whether temporal properties hold. Towards this end, we first provide a different characterisation of the satisfaction relation between a process and a formula in terms of game playing.

Exercises

1. **a.** Develop a set of goal directed rules for checking M properties of finite state processes. This can be viewed as a "top down" approach to property checking.

   b. Develop a "bottom up" approach to property checking of finite state processes. That is, develop a method that when given processes that have subformula properties, constructs the processes having the property itself.

   c. Give advantages and disadvantages of these two approaches, top down and bottom up. Is there a way of combining their advantages?

2. **a.** Develop a set of goal directed rules for checking CTL properties of finite state systems. Use your rules to show the following.

   i. $\texttt{Ven} \models \mathrm{AG}([\texttt{tick}]\texttt{ff} \vee \mathrm{AF}\langle \texttt{collect}_1 \rangle \texttt{tt})$

   ii. $\texttt{Cl} \models \mathrm{AG}(\langle - \rangle \texttt{tt} \wedge [-\texttt{tick}]\texttt{ff})$

   iii. $\texttt{Crossing} \models \mathrm{EF}\langle \texttt{ccross} \rangle \texttt{tt}$

   iv. $\texttt{Cl}_1 \models \mathrm{A}((\langle - \rangle \texttt{tt} \wedge [-\texttt{tick}]\texttt{ff})\,\mathrm{U}\,\langle \texttt{tock} \rangle \texttt{tt})$

   b. Develop a bottom up approach to CTL property checking of finite state processes. That is, develop a method that when given processes that have subformula properties, constructs the processes having the property itself. Illustrate your technique on the examples above.

3. Present an inductive technique for showing when $E \models_{\mathsf{V}} \nu Z.\, \Phi$ that uses induction on ordinals α.

6.2 Property checking games

We now present a game theoretic account of when a process E has a μM property Φ, relative to a valuation V. We assume that the formula Φ is normal, as defined in Section 5.2. Our intention is to present a game for property checking as we do for bisimulation checking in Chapter 3, so that a process has a property whenever player V has a winning strategy for the corresponding game.

- if $\Phi_j = \Psi_1 \wedge \Psi_2$, then player R chooses a conjunct Ψ_i where $i \in \{1, 2\}$: the process E_{j+1} is E_j and Φ_{j+1} is Ψ_i
- if $\Phi_j = \Psi_1 \vee \Psi_2$, then player V chooses a disjunct Ψ_i where $i \in \{1, 2\}$: the process E_{j+1} is E_j and Φ_{j+1} is Ψ_i
- if $\Phi_j = [K]\Psi$, then player R chooses a transition $E_j \xrightarrow{a} E_{j+1}$ with $a \in K$ and Φ_{j+1} is Ψ
- if $\Phi_j = \langle K \rangle \Psi$, then player V chooses a transition $E_j \xrightarrow{a} E_{j+1}$ with $a \in K$ and Φ_{j+1} is Ψ
- if $\Phi_j = \sigma Z.\, \Psi$, then Φ_{j+1} is Z and E_{j+1} is E_j
- if $\Phi_j = Z$ and the subformula of Φ_0 identified by Z is $\sigma Z.\, \Psi$, then Φ_{j+1} is Ψ and E_{j+1} is E_j

FIGURE 6.1. Rules for the next move in a game play

The "property checking game" $\mathsf{G}_\mathsf{V}(E, \Phi)$, when V is a valuation, E is a process and Φ is a normal formula, is played by two participants, players R (the refuter) and V (the verifier). Player R attempts to refute that $E \models_\mathsf{V} \Phi$, whereas player V wishes to establish that it is true.

A play of the game $\mathsf{G}_\mathsf{V}(E_0, \Phi_0)$ is a finite or infinite length sequence of the form

$$(E_0, \Phi_0) \ldots (E_n, \Phi_n) \ldots,$$

where each formula Φ_i is a subformula of Φ_0, that is $\Phi_i \in \mathrm{Sub}(\Phi_0)$, and each process E_i belongs to $\mathsf{P}(E_0)$. If part of a play is $(E_0, \Phi_0) \ldots (E_j, \Phi_j)$, then the next move, and which player makes it, depends on the main connective of the formula Φ_j. All the possibilities are presented in Figure 6.1. Players do not necessarily take turns, as in the bisimulation game[1]. There is a duality between the rules for $\wedge$ and $\vee$, and $[K]$ and $\langle K \rangle$. Player R makes the move when the main connective is $\wedge$ or $[K]$, whereas player V makes a similar move when it is $\vee$ or $\langle K \rangle$. The rules for fixed point formulas use the fact that the starting formula Φ_0 is normal, so each fixed point subformula is uniquely identified by its bound variable. Each time the current position is $(E, \sigma Z.\Psi)$, the next position is (E, Z), and each time it is (F, Z) the next position is (F, Ψ), meaning the fixed point subformula Z identifies is, in effect, unfolded once. Because there are no choices, neither player is responsible for these moves.

[1] It would be straightforward, but somewhat artificial, to make players take turn, by adding extra null moves.

Example 1 Consider the game $G(\texttt{Cl}, \nu Z.\,[\texttt{tock}]\texttt{ff} \wedge \langle\texttt{tick}\rangle Z)$[2]. The positions are as follows.

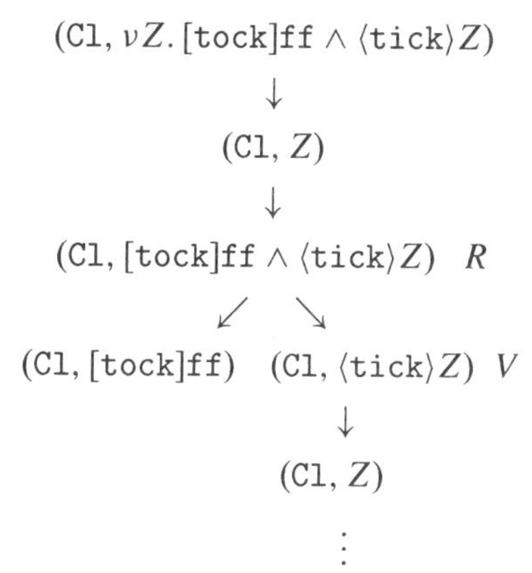

The arrows indicate which of the positions can lead to a subsequent position, and the label at a position indicates what player is responsible for the move. The position $(\texttt{Cl}, Z)$ is repeated. Hence, this game has only finitely many different positions. The "game graph" is the graphical presentation of the positions as above. A play of the game can be viewed as the sequence of positions that a token passes through as the players move it around the game graph. For instance, the single infinite play repeatedly cycles through $(\texttt{Cl}, Z)$.

Example 2 Consider the counter $\texttt{Ct}_0$ of Figure 1.4, which is a simple infinite state system and the game $G(\texttt{Ct}_0, \mu Z.\,[\texttt{up}]Z)$. The positions are very straightforward.

$$
\begin{array}{l}
(\texttt{Ct}_0, \mu Z.\,[\texttt{up}]Z) \\
\quad\downarrow \\
(\texttt{Ct}_0, Z) \\
\quad\downarrow \\
(\texttt{Ct}_0, [\texttt{up}]Z) \quad R \\
\quad\downarrow \\
(\texttt{Ct}_1, Z) \\
\quad\downarrow \\
\quad\vdots \\
\quad\downarrow \\
(\texttt{Ct}_i, Z) \\
\quad\downarrow \\
\quad\vdots
\end{array}
$$

[2] If a game does not depend on a valuation, we drop this index.

Player R wins

1. The play is $(E_0, \Phi_0) \ldots (E_n, \Phi_n)$ and
   - $\Phi_n = \texttt{ff}$, or
   - $\Phi_n = Z$ and Z is free in Φ_0 and $E_n \notin \mathsf{V}(Z)$, or
   - $\Phi_n = \langle K \rangle \Psi$ and $\{F : E_n \xrightarrow{a} F \text{ and } a \in K\} = \emptyset$
2. The play $(E_0, \Phi_0) \ldots (E_n, \Phi_n) \ldots$ has infinite length and the unique variable X, which occurs infinitely often and which subsumes all other variables occurring infinitely often, identifies a least fixed point subformula $\mu X.\, \Psi$

Player V wins

1. The play is $(E_0, \Phi_0) \ldots (E_n, \Phi_n)$ and
   - $\Phi_n = \texttt{tt}$, or
   - $\Phi_n = Z$ and Z is free in Φ_0 and $E_n \in \mathsf{V}(Z)$, or
   - $\Phi_n = [K]\Psi$ and $\{F : E_n \xrightarrow{a} F \text{ and } a \in K\} = \emptyset$
2. The play $(E_0, \Phi_0) \ldots (E_n, \Phi_n) \ldots$ has infinite length and the unique variable X, which occurs infinitely often and which subsumes all other variables occurring infinitely often, identifies a greatest fixed point subformula $\nu X.\, \Psi$

FIGURE 6.2. Winning conditions

The result is a game with an infinite number of different positions.

Next, we define when a player is said to win a play of a game. The winning conditions are presented in Figure 6.2. The refuter wins if a blatantly false position is reached, such as (E_n, Z), where Z is free in Φ_0 and $E_n \not\models_{\mathsf{V}} Z$. The verifier wins if a blatantly true position is reached. For instance, if the play reaches the position $(\texttt{Cl}, [\texttt{tock}]\texttt{ff})$ in example 1 because `Cl` has no `tock` transitions.

The second winning condition in Figure 6.2 identifies the winning player in an infinite length play. The winner depends on the "outermost fixed point" subformula that is unfolded infinitely often: if it is a least fixed point subformula, player R wins and if it is a greatest fixed point subformula, player V wins. Because there is only one fixed point subformula in the only infinite play of example 1, and it is a greatest fixed point, player V wins that play. Similarly, player R wins the only play of example 2 because the single fixed point variable Z identifies a least fixed point subformula. If there is more than one fixed point variable occurring infinitely often, then we need to know which of them is outermost. The definition in condition 2 is in terms of subsumption, as defined in definition 3 of Section 5.2. If X identifies $\sigma_1 X.\, \Psi$ and Z identifies $\sigma_2 Z.\, \Psi'$, then X subsumes Z if $\sigma_2 Z.\Psi' \in \mathrm{Sub}(\sigma_1 X.\Psi)$. In the case of a game involving a formula $\sigma_1 X_1 . \sigma_2 X_2 . \ldots \sigma_n X_n . \Phi(X_1, \ldots, X_n)$, any X_i may occur infinitely often in an infinite length play. However, there is just one X_j that occurs infinitely often and that subsumes any other X_k occurring infinitely often. This X_j is the outermost fixed point subformula and it decides which player wins the play. Proposition 1 makes precise this observation.

Proposition 1 *If $(E_0, \Phi_0) \ldots (E_n, \Phi_n) \ldots$ is an infinite length play of the game* $\mathsf{G_V}(E_0, \Phi_0)$, *then there is a unique variable X that*

1. *occurs infinitely often, that is for infinitely many j, $X = \Phi_j$, and*
2. *if Y also occurs infinitely often, then X subsumes Y.*

Proof. Let $\sigma X_1.\Psi_1, \ldots, \sigma X_n.\Psi_n$ be all the fixed point subformulas in $\mathrm{Sub}(\Phi_0)$ in decreasing order of size[3]. Therefore if $i < j$, then X_j cannot subsume X_i. Consider the next position (E_{j+1}, Φ_{j+1}) after (E_j, Φ_j) in a play: if Φ_j is not a variable, then $|\Phi_{j+1}| < |\Phi_j|$. Because each subformula has finite size, an infinite length play must proceed infinitely often through variables belonging to $\{X_1, \ldots, X_n\}$. Hence, there is at least one variable occurring infinitely often. If a subpart of the play has the form

$$(E_n, X_i) \ldots (E_k, X_j) \ldots (E_m, X_i) \ldots (E_l, X_j)$$

and $X_i \neq X_j$, then either X_i subsumes X_j or X_j subsumes X_i, but not both; see Proposition 1 of Section 5.2. Consequently, by transitivity of subsumption, there is exactly one variable X_i occurring infinitely often and subsuming any other X_j, which also occurs infinitely often. □

A strategy for a player is a family of rules telling the player how to move. It suffices (as with the bisimulation game of Chapter 3) to consider history-free strategies, whose rules do not depend upon previous positions in the play. For player R, rules have the following form.

- at position $(E, \Phi_1 \wedge \Phi_2)$ choose (E, Φ_i) where $i = 1$ or $i = 2$
- at position $(E, [K]\Phi)$ choose (F, Φ) where $E \xrightarrow{a} F$ and $a \in K$

For the verifier rules have a similar form.

- at position $(E, \Phi_1 \vee \Phi_2)$ choose (E, Φ_i) where $i = 1$ or $i = 2$
- at position $(E, \langle K \rangle \Phi)$ choose (F, Φ) where $E \xrightarrow{a} F$ and $a \in K$

A player uses the strategy π in a play provided all her moves in the play obey the rules in π. The strategy π is winning if the player wins every play in which she uses π. The following result provides an alternative account of the satisfaction relation between processes and formulas.

Theorem 1

1. $E \models_{\mathsf{V}} \Phi$ *iff player V has a history-free winning strategy for* $\mathsf{G_V}(E, \Phi)$.
2. $E \not\models_{\mathsf{V}} \Phi$ *iff player R has a history-free winning strategy for* $\mathsf{G_V}(E, \Phi)$.

The proof of Theorem 1 is deferred until the next section.

---

[3]The size of a formula Φ, written $|\Phi|$, is the number of connectives within it.

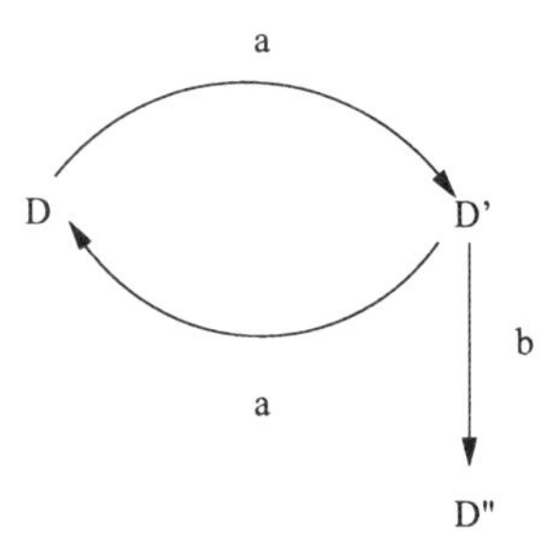

FIGURE 6.3. A simple process

Property checking games can be presented graphically, as in examples 1 and 2, where all the possible positions (those that are reachable in some play) are represented, together with which player is responsible for moving from a position, and where she is able to move to. Player V has a winning strategy for the game in example 1 and player R has a winning strategy for the game in example 2. In both cases the strategy consists of the empty set of rules (since the winning players have no choices).

Consider the simple process D in Figure 6.3, and the following formula Ψ.

$$\Psi = \mu Y.\, \nu Z.\, [a]((\langle b \rangle \mathtt{tt} \vee Y) \wedge Z)$$

D fails to have the property Ψ; see example 2 of Section 5.6. The refuter has a winning strategy for the game G(D, Ψ). The full game graph is pictured in Figure 6.4. The node labelled 16, (D, $\langle b \rangle \mathtt{tt}$), is a winning position for the refuter and node 9, (D$''$, $\mathtt{tt}$), is a winning position for the verifier. Player R's winning strategy consists of the following two rules.

at 14, (D, $(\langle b \rangle \mathtt{tt} \vee Y) \wedge Z$), choose 15, (D, $\langle b \rangle \mathtt{tt} \vee Y$)
at 6, (D$'$, $(\langle b \rangle \mathtt{tt} \vee Y) \wedge Z$), choose 12, (D$'$, Z)

Play will proceed from node 6 to node 12, and from node 14 to node 15. At node 15, player V either loses immediately by moving to 16 or returns to node 2. If player V always chooses node 2 when she is at 15, the play will be infinite. Although the two variables Z and Y occur infinitely often in this play (at nodes 2, 4 and 12), the refuter wins because Y subsumes Z and Y abbreviates a least fixed point formula.

In this graphical account, the positions represent the board of the game, and a token at a position represents the current position. A play is then a movement of the token around the board, with the responsible player at a position choosing the next position. An alternative presentation is to keep the process separate from the formula. Figure 6.5 is a picture of the above formula Ψ. Now we can think of a position as a pair of tokens, one that moves around the transition graph of the process, and the other that moves around the graph of the formula that is always finite. For example, the position (D, $(\langle b \rangle \mathtt{tt} \vee Y) \wedge Z$) would be represented with a token over D in Figure 6.3 and a token over 6 of Figure 6.5. It is the formula that

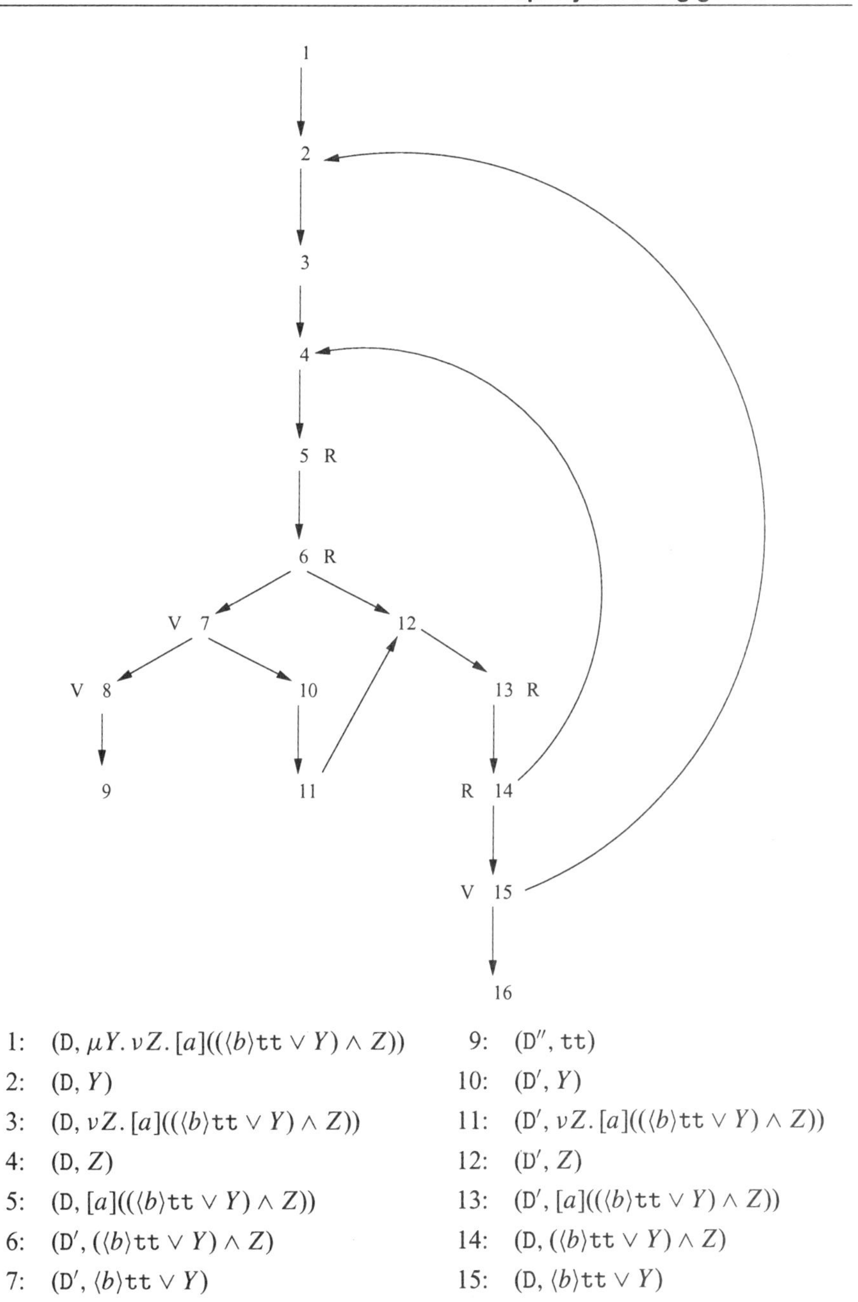

FIGURE 6.4. The game $\mathsf{G}(\mathtt{D}, \mu Y.\, \nu Z.\, [a]((\langle b\rangle \mathtt{tt} \vee Y) \wedge Z))$

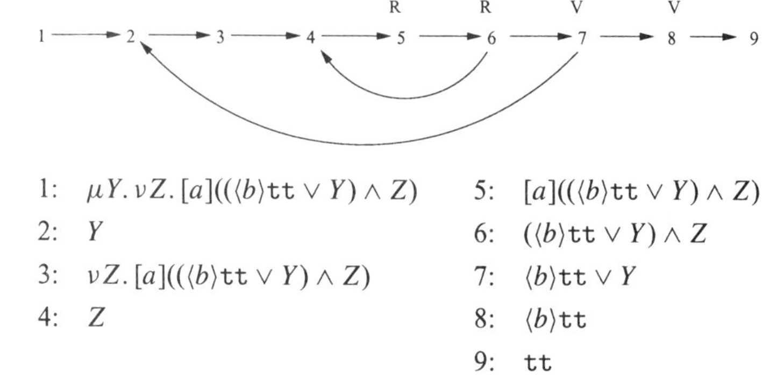

FIGURE 6.5. The formula $\mu Y.\, \nu Z.\, [a]((\langle b\rangle \mathtt{tt} \vee Y) \wedge Z)$

decides which player makes the next move. This is a more compact representation of a game because it requires only the sum of the size of the formula and the number of processes in a transition graph, whereas the representation using explicit positions may require the product of the two.

Example 3 The following family of processes is from example 2 of Section 5.7.

$$\mathrm{B}_0 \stackrel{\mathrm{def}}{=} a.\sum\{\mathrm{B}_i : i \geq 0\} + b.\sum\{\mathrm{B}_i : i \geq 0\}$$

$$\mathrm{B}_{i+1} \stackrel{\mathrm{def}}{=} b.\mathrm{B}_i \quad i \geq 0$$

For any $j \geq 0$, $\mathrm{B}_j \mid \mathrm{B}_j \not\models \mu Z.\, \langle -\rangle \mathtt{tt} \wedge [-a]Z$. This formula is pictured in Figure 6.6. Player R's winning strategy for any j is as follows.

at $\mathrm{B}_k \mid \mathrm{B}_j$ and formula 3 choose formula 5

at $\mathrm{B}_{k+1} \mid \mathrm{B}_j$ and formula 5 choose transition $\mathrm{B}_{k+1} \mid \mathrm{B}_j \stackrel{b}{\longrightarrow} \mathrm{B}_k \mid \mathrm{B}_j$

at $\mathrm{B}_0 \mid \mathrm{B}_j$ and formula 5 choose transition $\mathrm{B}_0 \mid \mathrm{B}_j \stackrel{b}{\longrightarrow} \mathrm{B}_j \mid \mathrm{B}_j$

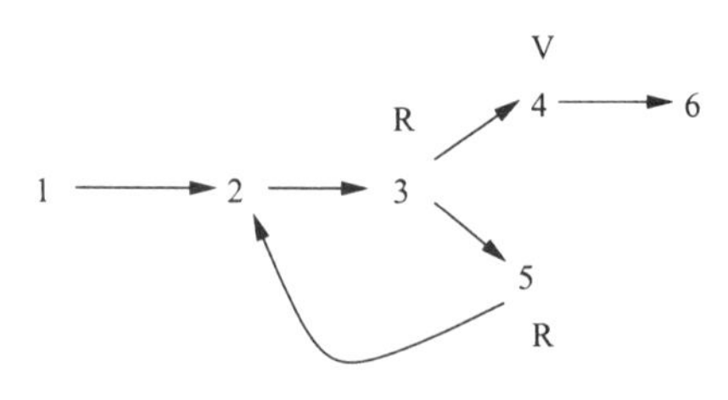

1: $\mu Z.\, \langle -\rangle \mathtt{tt} \wedge [-a]Z$
2: Z
3: $\langle -\rangle \mathtt{tt} \wedge [-a]Z$
4: $\langle -\rangle \mathtt{tt}$
5: $[-a]Z$
6: $\mathtt{tt}$

FIGURE 6.6. The formula $\mu Z.\, \langle -\rangle \mathtt{tt} \wedge [-a]Z$

For each j, the result is an infinite play passing through formula 2 infinitely often.

The game theoretic account of having a property holds for arbitrary processes whether they be finite or infinite state. Generally, property checking is undecidable: for instance, the halting problem is equivalent to whether $\mathtt{TM}_n \models \nu Z. \langle - \rangle \mathtt{tt} \wedge [-]Z$, where $\mathtt{TM}_n$ is the coding of the nth Turing machine. However, for some classes of infinite state processes property checking is decidable; see for example Esparza [21].

The general problem of property checking is closed under complement. Recall the definition of complement Φ^c of a formula Φ. For any valuation V, assume that V^c is its "complement" (with respect to a fixed E), the valuation such that for any Z, the set $\mathsf{V}^c(Z) = \mathsf{P}(E) - \mathsf{V}(Z)$. In the following result, let player P be either the refuter R or the verifier V throughout the Proposition.

Proposition 2 *Player P does not have a history-free winning strategy for* $\mathsf{G}_\mathsf{V}(E, \Phi)$ *iff Player P has a history-free winning stategy for* $\mathsf{G}_{\mathsf{V}^c}(E, \Phi^c)$.

Proof. $\mathsf{G}_{\mathsf{V}^c}(E, \Phi^c)$ is the dual of $\mathsf{G}_\mathsf{V}(E, \Phi)$, where the players reverse their roles, and the blatantly true and false configurations are interchanged. Therefore, if player P does not win $G_\mathsf{V}(E, \Phi)$, then the opponent player O wins this game. But then P wins the dual game $\mathsf{G}_{\mathsf{V}^c}(E, \Phi^c)$. The converse holds by similar reasoning. □

Exercises

1. Give winning strategies for the following properties of clocks
   - **a.** $\mathtt{Cl}_1 \models \nu Z. \langle \mathtt{tick} \rangle \langle \mathtt{tock} \rangle Z$
   - **b.** $\mathtt{Cl}_1 \models \mu Y. \langle \mathtt{tock} \rangle \mathtt{tt} \vee [-]Y$
   - **c.** $\mathtt{Cl}_2 \models \nu Z. [\mathtt{tick}, \mathtt{tock}]Z$
   - **d.** $\mathtt{Cl}_5 \models \nu Z. [\mathtt{tick}]Z$
   - **e.** $\mathtt{Cl}_6 \models \nu Z. \langle \tau \rangle Z$
   - **f.** $\mathtt{Cl}_7 \models \nu Z. \langle \mathtt{tick} \rangle \langle \mathtt{tick} \rangle Z \wedge \langle \mathtt{tock} \rangle \langle \mathtt{tock} \rangle Z$

   where $\mathtt{Cl}_1$, $\mathtt{Cl}_2$, and $\mathtt{Cl}_5$ are as in Section 1.1, $\mathtt{Cl}_6 \stackrel{\mathrm{def}}{=} \mathtt{tick}.\mathtt{Cl}_6 + \tau.\mathtt{Cl}_6$, and $\mathtt{Cl}_7 \stackrel{\mathrm{def}}{=} \mathtt{tick}.\mathtt{Cl}_7 + \mathtt{tock}.\mathtt{Cl}_7$.
2. Show the following using games
   - **a.** $\mathtt{Cl}_1 \models \mathrm{Cycle}(\mathtt{tick}\ \mathtt{tock})$
   - **b.** $\mathtt{Sched}'_3 \models \mathrm{Cycle}(a_1\, a_2\, a_3)$
   - **c.** $\mathtt{Sched}_3 \models \mathrm{Cycle}(a_1\, a_2\, a_3)$

   where the property Cycle(...) is from Section 5.7, and the schedulers are from Section 1.4.
3. Show the following using games
   - **a.** $\mathtt{T}(17) \vdash \mu Y. \langle - \rangle \mathtt{tt} \wedge [-\overline{\mathtt{out}}(1)]Y$
   - **b.** $\mathtt{Sem} \mid \mathtt{Sem} \mid \mathtt{Sem} \vdash \nu Z. [\mathtt{get}](\mu Y. \langle - \rangle \mathtt{tt} \wedge [-\mathtt{put}]Y) \wedge [-]Z$

where $\mathtt{T}(i)$ is defined in Section 1.1, and $\mathtt{Sem} \stackrel{\text{def}}{=} \mathtt{get.put.Sem}$.

4. Using games, show $\mathtt{A}_j \mid \mathtt{A}_j \models \mu Z.\, \langle - \rangle \mathtt{tt} \wedge [-a]Z$, for any $j \geq 0$, where $\mathtt{A}_j$ is defined in example 2 of Section 5.7.
5. Define complement games for the following.
   - **a.** $\mathtt{Cl}_1 \models \nu Z.\, \langle \mathtt{tick} \rangle \langle \mathtt{tock} \rangle Z$
   - **b.** $\mathtt{Cl}_1 \models \mu Y.\, \langle \mathtt{tock} \rangle \mathtt{tt} \vee [-]Y$
   - **c.** $\mathtt{Cl}_1 \models \text{Cycle}(\mathtt{tick}\ \mathtt{tock})$

6.3 Correctness of games

This section is devoted to the proof of Theorem 1 of the previous section, which shows that games provide an alternative basis for processes having, or failing to have, a property. The result appeals to approximants, and uses an extension of the "least approximant" result; Proposition 2 of Section 5.5.

Theorem 1

1. $E \models_{\mathsf{V}} \Phi$ *iff player V has a history-free winning strategy for* $\mathsf{G}_{\mathsf{V}}(E, \Phi)$.
2. $E \not\models_{\mathsf{V}} \Phi$ *iff player R has a history-free winning strategy for* $\mathsf{G}_{\mathsf{V}}(E, \Phi)$.

Proof. Assume that $E_0 \models_{\mathsf{V}} \Phi_0$. We show that player V has a history-free winning strategy for $\mathsf{G}_{\mathsf{V}}(E_0, \Phi_0)$. The proof idea is straightforward, since we show that player V is always able to preserve "truth" of game configurations by making her choices wisely, so winning any play. The subtlety is the construction of the history-free winning strategy, and it is here that we use approximants to make optimal choices. We develop first the idea of "least" approximants.

Let $\sigma_1 Z_1.\Psi_1, \ldots, \sigma_n Z_n.\Psi_n$ be all the fixed point subformulas in $\text{Sub}(\Phi_0)$ in decreasing order of size (as in the proof of Lemma 1 of the previous section). Therefore, if Z_i subsumes Z_j, this means that $i \leq j$: it is not possible for Z_j to subsume Z_i when $i < j$. A position in the game $\mathsf{G}_{\mathsf{V}}(E_0, \Phi_0)$ has the form (F, Ψ), where Ψ may contain free variables in $\{Z_1, \ldots, Z_n\}$. But really these variables are bound. Therefore, we now consider the correct semantics for understanding the formulas in a play.

Let P be the set of processes $\mathsf{P}(E_0)$. We define valuations $\mathsf{V}_0, \ldots, \mathsf{V}_n$ iteratively as follows

$$\begin{aligned} \mathsf{V}_0 &= \mathsf{V} \\ \mathsf{V}_{i+1} &= \mathsf{V}_i[\mathsf{E}_{i+1}/Z_{i+1}], \end{aligned}$$

where $\mathsf{E}_{i+1} = \{E \in \mathsf{P} : E \models_{\mathsf{V}_i} \sigma_{i+1} Z_{i+1} \,.\, \Psi_{i+1}\}$. The valuation V_n captures the meaning of all the bound variables Z_i. We say that a game position (F, Ψ) is true if $F \models_{\mathsf{V}_n} \Psi$. Clearly, the starting position is true because $E_0 \models_{\mathsf{V}_n} \Phi_0$ iff $E_0 \models_{\mathsf{V}} \Phi_0$, and by assumption $E_0 \models_{\mathsf{V}} \Phi_0$.

Given any true position, we define a refined valuation that identifies the smallest least fixed point approximants making the configuration true. To define this, let $\mu Y_1 . \Psi'_1, \ldots, \mu Y_k . \Psi'_k$ be all least fixed point subformulas in $\mathrm{Sub}(\Phi_0)$ in decreasing order of size, meaning each Y_i is some Z_j. We define a signature as a sequence of ordinals of length $k, \alpha_1 \ldots \alpha_k$. We assume the lexicographic ordering on signatures: $\alpha_1 \ldots \alpha_k < \beta_1 \ldots \beta_k$ if there is an $i \leq k$ such that $\alpha_j = \beta_j$ for all $1 \leq j < i$ and $\alpha_i < \beta_i$. Given a signature $s = \alpha_1 \ldots \alpha_k$, we define its associated valuation V^s_n as follows. Again the definition is iterative, since we define valuations $\mathsf{V}^s_0, \ldots, \mathsf{V}^s_n$

$$
\begin{aligned}
\mathsf{V}^s_0 &= \mathsf{V} \\
\mathsf{V}^s_{i+1} &= \mathsf{V}^s_i[\mathsf{E}_{i+1}/Z_{i+1}],
\end{aligned}
$$

where now the set E_{i+1} depends on the kind of fixed point $\sigma_{i+1} Z_{i+1}$.

1. If $\sigma_{i+1} = \nu$ then $\mathsf{E}_{i+1} = \{E \in \mathsf{P} : E \models_{\mathsf{V}^s_i} \sigma_{i+1} Z_{i+1} . \Psi_{i+1}\}$
2. If $\sigma_{i+1} Z_{i+1} = \mu Y_j$ then $\mathsf{E}_{i+1} = \{E \in \mathsf{P} : E \models_{\mathsf{V}^s_i} \sigma Y^{\alpha_j}_j . \Psi'_j\}$

We leave as an exercise that, if $F \models_{\mathsf{V}_n} \Psi$, then there is indeed a smallest signature s such that $F \models_{\mathsf{V}^s_n} \Psi$. It is this result that generalises the "least approximant" result of Section 5.5. Given a true configuration (F, Ψ), we therefore define its signature to be the least s such that $F \models_{\mathsf{V}^s_n} \Psi$. Signatures were introduced by Streett and Emerson in [56].

Next, we define the history-free winning strategy for player V. Consider any true position, which is a player V position. If the position is $(F, \Psi_1 \vee \Psi_2)$, then consider its signature s. It follows that $F \models_{\mathsf{V}^s_n} \Psi_1 \vee \Psi_2$. Therefore, $F \models_{\mathsf{V}^s_n} \Psi_i$ for $i = 1$ or $i = 2$. Choose one of these Ψ_i that holds, and add the rule "at $(F, \Psi_1 \vee \Psi_2)$ choose (F, Ψ_i)." The construction is similar for a true position of the form $(F, \langle K \rangle \Psi)$. Let s be its signature, and therefore $F \models_{\mathsf{V}^s_n} \langle K \rangle \Psi$. Hence, there is a transition $F \xrightarrow{a} F'$ with $a \in K$, and $F' \models_{\mathsf{V}^s_n} \Psi$. Therefore, the rule in player V's strategy is "at position $(F, \langle K \rangle \Psi)$ choose (F', Ψ)." The result is a history-free strategy. The rest of the proof consists of showing that indeed it is a winning strategy.

But suppose not. Assume that player R can defeat player V. The starting position is true, and as play proceeds we show that the play can not reach a false position because player V uses her strategy. Moreover, we shall carry around the signature of each position. The initial position is (E_0, Φ_0) and $E_0 \models_{\mathsf{V}^{s_0}_n} \Phi_0$. Assume that (E_m, Φ_m) is the current position in the play and $E_m \models_{\mathsf{V}^{s_m}_n} \Phi_m$. If the position is a final position, then clearly player R is not the winner. Therefore, either player V wins, or the play is not yet complete. In the latter case, we show how the play is extended to (E_{m+1}, Φ_{m+1}) by case analysis on Φ_m.

If $\Phi_m = \Psi_1 \wedge \Psi_2$, then player R chooses Ψ_i, $i \in \{1, 2\}$, and the next position (E_{m+1}, Φ_{m+1}) is true where $E_{m+1} = E_m$ and $\Phi_{m+1} = \Psi_i$, and $E_{m+1} \models_{\mathsf{V}^{s_{m+1}}_n} \Phi_{m+1}$. Clearly, the signature $s_{m+1} \leq s_m$. A similar argument applies to the other player R move when $\Phi_m = [K]\Psi$, and again $s_{m+1} \leq s_m$.

If $\Phi_m = \Psi_1 \vee \Psi_2$, then player V uses the strategy to choose the next position. Clearly, this preserves truth because s_m must be the same signature as when defining the strategy. A similar argument applies to the other case of a player V move when $\Phi_m = \langle K \rangle \Psi$. Again, in both cases $s_{m+1} \leq s_m$.

If $\Phi_m = \sigma_i Z_i . \Psi_i$, then the next game configuration is the true position (E_{m+1}, Φ_{m+1}), where $E_{m+1} = E_m$ and $\Phi_{m+1} = Z_i$. Notice the signature may increase if $\sigma_i = \mu$. If $\Phi_m = Z_i$ and $\sigma_i = \nu$, then the next position (E_{m+1}, Φ_{m+1}) is true, where $E_{m+1} = E_m$ and $\Phi_{m+1} = \Psi_i$ and $s_{m+1} = s_m$. If $\Phi_m = Z_i$ and $\sigma_i = \mu$ then the next position (E_{m+1}, Φ_{m+1}) is true, where $E_{m+1} = E_m$ and $\Phi_{m+1} = \Psi_i$ and $s_{m+1} < s_m$ because, in effect, the fixed point has been unfolded. In this case there is a genuine decrease in signature.

The proof is completed by showing that player V must win any such play. As we have seen, this is the case when a play has finite length. Consider therefore an infinite length play. By Proposition 1 of the previous section, there is a unique Z_i occurring infinitely often, and which subsumes any other Z_j also occurring infinitely often. Consider the game play $(E_k, \Phi_k) \ldots$ from the point k such that every occurrence of a variable Z_j in the play is subsumed by Z_i, and let $k1$, $k2$, ... be the positions in this suffix play where Z_i occurs. Z_i cannot identify a least fixed point subformula. For by the construction above, this would require there to be a strictly decreasing sequence of signatures sets $s_{k1} > s_{k2} > \ldots$, which is impossible. Therefore Z_i identifies a greatest fixed point subformula.

Part 2 of the proof is similar. We need to generalise the other half of the least approximant result; Proposition 2 of Section 5.5. The proof is left as an exercise for the reader. □

Exercises

1. Prove the smallest signature result used in Theorem 1: if $F \models_{V_n} \Psi$, then there is a smallest signature s such that $F \models_{V_n^s} \Psi$.
2. For each of the following construct its signature

   a. $\texttt{Cl}_1 \models \nu Z. \langle \texttt{tick} \rangle \langle \texttt{tock} \rangle Z$

   b. $\texttt{Cl}_1 \models \mu Y. \langle \texttt{tock} \rangle \texttt{tt} \vee [-] Y$

   c. $\texttt{T(19)} \vdash \mu Y. \langle - \rangle \texttt{tt} \wedge [-\overline{\texttt{out}}(1)] Y$

   d. $\texttt{Sem} \mid \texttt{Sem} \mid \texttt{Sem} \vdash \nu Z. [\texttt{get}](\mu Y. \langle - \rangle \texttt{tt} \wedge [-\texttt{put}] Y) \wedge [-] Z$

   e. $\texttt{A}_5 \mid \texttt{A}_5 \models \mu Z. \langle - \rangle \texttt{tt} \wedge [-a] Z$

   where $\texttt{A}_5$ is defined in example 2 of Section 5.7.
3. Prove part 2 of Theorem 1.

6.4 CTL games

The two until operators of the logic CTL of Section 4.3 can be viewed as macros for μM formulas, as described in Section 5.2.

$$\begin{array}{lcl} \mathrm{A}(\Phi\,\mathrm{U}\,\Psi) & \equiv & \mu Z.\,\Psi \vee (\Phi \wedge (\langle - \rangle \mathtt{tt} \wedge [-]Z)) \\ \mathrm{E}(\Phi\,\mathrm{U}\,\Psi) & \equiv & \mu Z.\,\Psi \vee (\Phi \wedge \langle - \rangle Z) \end{array}$$

Negations of until formulas are therefore maximal fixed point formulas, as also described in Section 5.2. In this section, property checking games for CTL are presented. The idea is to develop CTL games entirely within the syntax of CTL, even though their justification owes to the fact that CTL can be embedded in μMI, as shown in Section 5.6. Similar games could be developed for any other logic that can be systematically defined in μM, such as propositional dynamic logic.

The syntax of CTL is given in Section 4.3. In the following, for ease of exposition, the following abbreviation is used.

$$\langle K \rangle \Psi \equiv \neg [K] \neg \Psi$$

A play of the CTL property checking game $\mathsf{G}(E_0, \Phi_0)$, where Φ_0 is a CTL formula, is a finite or an infinite length sequence of pairs (E_i, Φ_i). The next move in the play $(E_0, \Phi_0) \ldots (E_j, \Phi_j)$ is defined in Figure 6.7. The rules appear to be complicated, but this is because of the presence of negation in the logic. The refuter

- if $\Phi_j = \Psi_1 \wedge \Psi_2$, then player R chooses a conjunct Ψ_i, where $i \in \{1, 2\}$: the process E_{j+1} is E_j and Φ_{j+1} is Ψ_i
- if $\Phi_j = \neg(\Psi_1 \wedge \Psi_2)$, then player V chooses a conjunct Ψ_i, where $i \in \{1, 2\}$: the process E_{j+1} is E_j and Φ_{j+1} is $\neg\Psi_i$
- if $\Phi_j = [K]\Psi$, then player R chooses a transition $E_j \xrightarrow{a} E_{j+1}$ with $a \in K$ and Φ_{j+1} is Ψ
- if $\Phi_j = \neg[K]\Psi$, then player V chooses a transition $E_j \xrightarrow{a} E_{j+1}$ with $a \in K$ and Φ_{j+1} is $\neg\Psi$
- if $\Phi_j = \neg\neg\Psi$, then E_{j+1} is E_j and Φ_{j+1} is Ψ
- if $\Phi_j = \mathrm{A}(\Psi_1 \mathrm{U} \Psi_2)$, then $\neg(\neg\Psi_2 \wedge \neg(\Psi_1 \wedge (\langle - \rangle \mathtt{tt} \wedge [-]\mathrm{A}(\Psi_1 \mathrm{U} \Psi_2))))$ is Φ_{j+1} and E_{j+1} is E_j
- if $\Phi_j = \neg\mathrm{A}(\Psi_1 \mathrm{U} \Psi_2)$, then $\neg\Psi_2 \wedge \neg(\Psi_1 \wedge (\langle - \rangle \mathtt{tt} \wedge [-]\mathrm{A}(\Psi_1 \mathrm{U} \Psi_2)))$ is Φ_{j+1} and E_{j+1} is E_j
- if $\Phi_j = \mathrm{E}(\Psi_1 \mathrm{U} \Psi_2)$, then $\neg(\neg\Psi_2 \wedge \neg(\Psi_1 \wedge \langle - \rangle \mathrm{E}(\Psi_1 \mathrm{U} \Psi_2)))$ is Φ_{j+1} and E_{j+1} is E_j
- if $\Phi_j = \neg\mathrm{E}(\Psi_1 \mathrm{U} \Psi_2)$, then $\neg\Psi_2 \wedge \neg(\Psi_1 \wedge \langle - \rangle \mathrm{E}(\Psi_1 \mathrm{U} \Psi_2))$ is Φ_{j+1} and E_{j+1} is E_j

FIGURE 6.7. Rules for the next move in a CTL game play

Player R wins

1. The play is $(E_0, \Phi_0) \ldots (E_n, \Phi_n)$ and
   - $\Phi_n = \neg\texttt{tt}$, or
   - $\Phi_n = \neg[K]\Psi$ and $\{F : E \xrightarrow{a} F \text{ and } a \in K\} = \emptyset$
2. The play $(E_0, \Phi_0) \ldots (E_n, \Phi_n) \ldots$ has infinite length and there is an until formula $\mathrm{A}(\Psi_1 \mathrm{U} \Psi_2)$ or $\mathrm{E}(\Psi_1 \mathrm{U} \Psi_2)$ occurring infinitely often in the play

Player V wins

1. The play is $(E_0, \Phi_0) \ldots (E_n, \Phi_n)$ and
   - $\Phi_n = \texttt{tt}$, or
   - $\Phi_n = [K]\Psi$ and $\{F : E \xrightarrow{a} F \text{ and } a \in K\} = \emptyset$
2. The play $(E_0, \Phi_0) \ldots (E_n, \Phi_n) \ldots$ has infinite length and there is a negated until formula $\neg\mathrm{A}(\Psi_1 \mathrm{U} \Psi_2)$ or $\neg\mathrm{E}(\Psi_1 \mathrm{U} \Psi_2)$ occurring infinitely often in the play

FIGURE 6.8. Winning conditions

chooses the next position when the formula has the form $\Psi_1 \wedge \Psi_2$ or $[K]\Psi$, and the verifier chooses when it has the form $\neg(\Psi_1 \wedge \Psi_2)$ or $\neg[K]\Psi$. Because there are no choices in the remaining rules, neither player is responsible for them. The first of these reduces a double negation. The remaining rules are for until formulas and their negations, and are determined by their fixed point definitions.

Figure 6.8 captures when a player is said to win a play of a game. Player R wins a play if a blatantly false position is reached, and V wins if an obviously true position is reached. The other condition identifies which of the players wins an infinite length play. This is much easier to decide than for full μM. For any infinite length play of a CTL game, there is only one until formula or negation of an until formula occurring infinitely often. It is this formula that decides who wins. If it is an until formula, and therefore a least fixed point formula, then R wins the play; if it is the negation of an until formula, and therefore a maximal fixed point formula, then V wins the play.

Again a history-free strategy for a player is a family of rules telling the player how to move, and independent of previous positions. For the refuter, rules have the following form.

- at position $(E, \Phi_1 \wedge \Phi_2)$ choose (E, Φ_i) where $i = 1$ or $i = 2$
- at position $(E, [K]\Phi)$ choose (F, Φ) where $E \xrightarrow{a} F$ and $a \in K$

For the verifier rules have a similar form.

- at position $(E, \neg(\Phi_1 \wedge \Phi_2))$ choose $(E, \neg\Phi_i)$ where $i = 1$ or $i = 2$
- at position $(E, \neg[K]\Phi)$ choose $(F, \neg\Phi)$ where $E \xrightarrow{a} F$ and $a \in K$

A player uses the strategy π in a play if all her moves in the play obey the rules in π, and π is winning if the player wins every play in which she uses π.

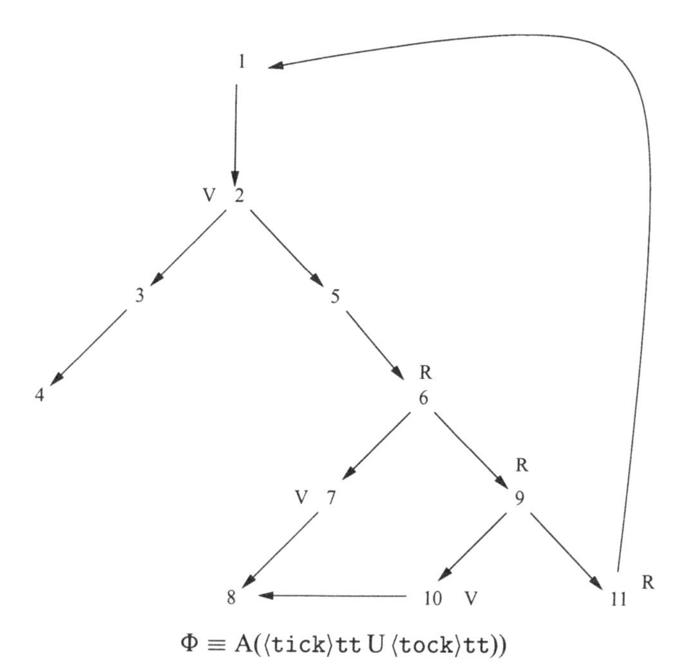

$$
\begin{aligned}
1 &: (\mathtt{Cl}, \Phi)\\
2 &: (\mathtt{Cl}, \neg(\neg\langle\mathtt{tock}\rangle\mathtt{tt} \wedge \neg(\langle\mathtt{tick}\rangle\mathtt{tt} \wedge (\langle-\rangle\mathtt{tt} \wedge [-]\Phi))))\\
3 &: (\mathtt{Cl}, \neg\neg\langle\mathtt{tock}\rangle\mathtt{tt})\\
4 &: (\mathtt{Cl}, \langle\mathtt{tock}\rangle\mathtt{tt})\\
5 &: (\mathtt{Cl}, \neg\neg(\langle\mathtt{tick}\rangle\mathtt{tt} \wedge (\langle-\rangle\mathtt{tt} \wedge [-]\Phi)))\\
6 &: (\mathtt{Cl}, \langle\mathtt{tick}\rangle\mathtt{tt} \wedge (\langle-\rangle\mathtt{tt} \wedge [-]\Phi))\\
7 &: (\mathtt{Cl}, \langle\mathtt{tick}\rangle\mathtt{tt})\\
8 &: (\mathtt{Cl}, \mathtt{tt})\\
9 &: (\mathtt{Cl}, \langle-\rangle\mathtt{tt} \wedge [-]\Phi)\\
10 &: (\mathtt{Cl}, \langle-\rangle\mathtt{tt})\\
11 &: (\mathtt{Cl}, [-]\Phi)
\end{aligned}
$$

FIGURE 6.9. A CTL game

Example 1 Just for illustration, we show that the refuter has a winning strategy for the game $\mathsf{G}(\mathtt{Cl}, \mathrm{A}(\langle\mathtt{tick}\rangle\mathtt{tt}\mathrm{U}\langle\mathtt{tock}\rangle\mathtt{tt}))$. The game graph is presented in Figure 6.9. Position 4 is a winning position for the refuter, and 8 is a winning position for the verifier. The refuter's winning strategy is to choose 9 at 6, and 11 at 9. Therefore, the only infinite play cycles through the until formula, and is thereby won by player R.

The following Proposition is a corollary of Theorem 1 of Section 6.3.

Proposition 1

1. *If* $\Phi \in$ CTL, *then* $E \models \Phi$ *iff player* V *has a history-free winning strategy for* $\mathsf{G}(E, \Phi)$.
2. *If* $\Phi \in$ CTL, *then* $E \not\models \Phi$ *iff player* R *has a history-free winning strategy for* $\mathsf{G}(E, \Phi)$.

The characteristic of a play of a CTL game is inherited for any play of a game $\mathsf{G}_\mathsf{V}(E, \Phi)$ when $\Phi \in \mu$MI, that for any infinite play there is just one fixed point variable that occurs infinitely often. A more general fragment of μM is μMA, the alternation free formulas of modal mu-calculus. These fragments are defined in Section 5.6. In any infinite play of $\mathsf{G}_\mathsf{V}(E, \Phi)$ when $\Phi \in \mu$MA all fixed point variables occurring infinitely often are of the same kind, that is, they are either all greatest or all least fixed point variables.

Proposition 2

1. *If* $\Phi \in \mu$MI, *then in any infinite play of the game* $\mathsf{G}_\mathsf{V}(E, \Phi)$ *there is just one fixed point variable* X *occurring infinitely often.*
2. *If* $\Phi \in \mu$MA, *then in any infinite play of the game* $\mathsf{G}_\mathsf{V}(E, \Phi)$ *the variables occurring infinitely often are either all greatest fixed point variables, or are all least fixed point variables.*

Proof. For both cases it is clear that, in any infinite length play, there is at least one fixed point variable that occurs infinitely often. If there is just one fixed point variable that occurs infinitely often, then the results follow. Otherwise, assume an infinite play in which the distinct variables $X_1, \ldots, X_n$ all occur infinitely often. Assume that they identify the fixed point subformulas $\sigma_1 X_1 . \Psi_1, \ldots, \sigma_n X_n . \Psi_n$, in decreasing order of size. Consider now the first case, when $\Phi \in \mu$MI. Because it is possible to proceed in a number of moves from some position (F_1, X_1) to (F_2, X_2), and then to (F_3, X_3) and so on to (F_n, X_n) and then back to (F'_1, X_1), it follows that X_1 must occur free in at least one of the subformulas $\sigma_2 X_1 . \Psi_1, \ldots, \sigma_n X_n . \Psi_n$, which in the first case contradicts that the starting formula $\Phi \in \mu$MI. For the second case, assume that X_i is the first variable that identifies a different kind of fixed point to that of X_1. Because it is possible to proceed in a number of moves from (F_i, X_i) to (F'_1, X_1), at least one of the variables X_j, $1 \leq j < i$ must occur free in $\sigma_i X_i . \Psi_i$, which contradicts that the starting formula $\Phi \in \mu$MA. □

Exercises

1. Give winning strategies for the following properties of `Ven`
   a. `Ven` $\models$ A(`tt` U ⟨`collect`$_\mathtt{b}$, `collect`$_\mathtt{l}$⟩`tt`)
   b. `Ven` $\models$ E(`tt` U ⟨`collect`$_\mathtt{b}$⟩`tt`)
   c. `Ven` $\models$ E(`tt` U ⟨`collect`$_\mathtt{l}$⟩`tt`)

d. $\texttt{Ven} \models \neg A(\texttt{tt}\, U\, \langle \texttt{collect}_1 \rangle \texttt{tt})$

2. Extend the rules of CTL game playing to formulas of the form $A\,F\Phi$, $E\,F\Phi$. Give the winning strategies for the following.

   a. $\texttt{SM}_0 \models E\,F\langle \overline{\texttt{win}}(5) \rangle \texttt{tt}$

   b. $\texttt{Crossing} \models \neg E\,F(\langle \overline{\texttt{tcross}} \rangle \texttt{tt} \wedge \langle \overline{\texttt{ccross}} \rangle \texttt{tt})$

3. Consider AD_n as defined in the exercises in Section 5.6. Give a condition, similar in spirit to Proposition 2, that characterises how many different alternations of fixed point variables any infinite play of a game involving $\Phi \in \text{AD}_n$ can have.

6.5 Parity games

Assume that E is a finite state process. An algorithm answering the question "is it the case that $E \models_V \Phi$?" is known as a model checker. Process E determines a model, its transition graph, and the question is whether the property Φ holds at state E of this model (relative to V). Model checking was introduced by Clarke, Emerson and Sistla for the logic CTL [12]. Games offer a foundation for model checking. However, there are other foundations including nondeterministic automata and alternating automata; see Vardi and Wolper, and Bernholtz et al., [59, 4].

An important issue is how much resource is needed to model check. There are questions of time (How quickly can it be done?) and space (How much memory is needed?). The quantitative results depend on what counts as the size of the decision problem. One notion of size is that of the resulting game graph, whose upper bound is $|\mathsf{P}(E)| \times |\Phi|$. Alternatively, because the input to model checking is the model E and the formula Φ, the size is $|\mathsf{P}(E)| + |\Phi|$. The contrast between these two notions of size is illustrated by the two graphical presentations of the property checking game in Section 6.3. Both notions of size depend on the model of E as a transition graph. A third notion of size does not appeal to this graph. Instead, the size of E is the size of its description in CCS. For instance, if E is a parallel composition of processes $E_1 | \ldots | E_n$ then the size of its description is the sum of the sizes of the components, whereas the size of its transition graph is the product of the sizes of its components.

Whatever the notion of size, there is also the question of an efficient implementation of the decision question. What succinct data structures should be used for representing processes, their behaviour, and formulas? One popular method is to represent graphs succinctly using OBDDs (ordered binary decision diagrams).

Model checking provides a "yes" or "no" answer to the question, "is it the case that $E \models_V \Phi$?" A more informative answer includes why E satisfies or fails to satisfy Φ. The property checking game offers the possibility of exhibiting the winning strategy that can then be used as evidence.

In this section, model checking is abstracted into a simpler graph game, (which is a slight variant of what is called the parity game; see Emerson and Jutla [19]). A parity game is a directed graph $\mathsf{G} = (N, \rightarrow, L)$ whose set of vertices N is a finite subset of $\mathbb{N}$ and whose binary edge relation $\rightarrow$ relates vertices. As usual we write $i \rightarrow j$ instead of $(i, j) \in \rightarrow$. The vertices are the positions of the game. The third component L labels each vertex with R or with V: $L(i)$ tells us which player is responsible for moving from vertex i. A play always has infinite length because we impose the condition that each vertex i has at least one edge $i \longrightarrow j$. A parity game is a contest between player R and V. It begins with a token on the least vertex j (with respect to $<$ on $\mathbb{N}$). When the token is on vertex i and $L(i) = P$, player P moves it along one of the outgoing edges of i. A play therefore consists of an infinite length path through the graph along which the token passes.

The winner of a play is determined by the label of the *least* vertex i that occurs infinitely often in the play. If $L(i) = R$, then player R wins; otherwise $L(i) = V$, and player V wins. We now show how to transform the property checking game for finite state processes into an equivalent parity game. Let $\mathsf{G}_\mathsf{V}(E, \Phi)$ be the property checking game for the finite state process E and the normal formula Φ. The transformation into the parity game $\mathsf{G}_\mathsf{V}[E, \Phi]$ proceeds as follows.

1. Let $E_1, \ldots, E_m$ be a list of all processes in $\mathsf{P}(E)$ with $E = E_1$.
2. Let $Z_1, \ldots, Z_k$ be a list of all bound variables in Φ. Therefore, for each Z_i there is the associated subformula $\sigma_i Z_i.\Psi_i$.
3. Let $\Phi_1, \ldots, \Phi_l$ be a list of all formulas in $\mathrm{Sub}(\Phi) - \{Z_1, \ldots, Z_k\}$ in decreasing order of size. This means that $\Phi = \Phi_1$. Extend the list by inserting each Z_i directly after the fixed point subformula associated with it: $\Phi_1, \ldots, \sigma_i Z_i.\Psi_i, Z_i, \ldots, \Phi_l$. The result is a list of all formulas $\Phi_1, \ldots, \Phi_n$ in $\mathrm{Sub}(\Phi)$ in decreasing order of size, except that the bound variable Z_i is "bigger" than Ψ_i but "smaller" than $\sigma_i Z_i.\Psi_i$.
4. The possible positions of the property checking game $\mathsf{G}_\mathsf{V}(E, \Phi)$ is now listed in the following order. Positions earlier in the list contain "larger" formulas.

   $$(E_1, \Phi_1), \ldots, (E_m, \Phi_1), (E_1, \Phi_2), \ldots, (E_1, \Phi_n), \ldots, (E_m, \Phi_n)$$

   The vertex set N of the parity game $\mathsf{G}_\mathsf{V}[E, \Phi]$ is $\{1, \ldots, m \times n\}$. Each vertex $i = m \times (k - 1) + j$ represents the position (E_j, Φ_k).
5. Next, we define the labelling $L(i)$ of vertex i and its edges by case analysis on the position (F, Ψ) that i represents.
   - If Ψ is Z and Z is free in the starting formula Φ and $F \in \mathsf{V}(Z)$, then $L(i) = V$ and there is the edge $i \rightarrow i$. Instead, if $F \notin \mathsf{V}(Z)$, then $L(i) = R$ and there is the edge $i \rightarrow i$
   - If Ψ is `tt`, then $L(i) = V$ and there is the edge $i \rightarrow i$
   - If Ψ is `ff`, then $L(i) = R$ and there is the edge $i \rightarrow i$
   - If Ψ is $\Psi_1 \wedge \Psi_2$, then $L(i) = R$ and there are edges $i \rightarrow j1$ and $i \rightarrow j2$ where $j1$ represents (F, Ψ_1) and $j2$ represents (F, Ψ_2)

- If Ψ is $\Psi_1 \vee \Psi_2$, then $L(i) = V$ and there are edges $i \rightarrow j1$ and $i \rightarrow j2$ where $j1$ represents (F, Ψ_1) and $j2$ represents (F, Ψ_2)
- If Ψ is $[K]\Psi'$ and $\{F' : F \xrightarrow{a} F' \text{ and } a \in K\} = \emptyset$, then $L(i) = V$ and there is the edge $i \rightarrow i$
- If Ψ is $[K]\Psi'$ and $\{F' : F \xrightarrow{a} F' \text{ and } a \in K\} \neq \emptyset$, then $L(i) = R$ and there is an edge $i \rightarrow j$ for each j representing a position (F', Ψ') such that $F \xrightarrow{a} F'$ for $a \in K$
- If Ψ is $\langle K\rangle\Psi'$ and $\{F' : F \xrightarrow{a} F' \text{ and } a \in K\} = \emptyset$, then $L(i) = R$ and there is the edge $i \rightarrow i$
- If Ψ is $\langle K\rangle\Psi'$ and $\{F' : F \xrightarrow{a} F' \text{ and } a \in K\} \neq \emptyset$, then $L(i) = V$ and there is an edge $i \rightarrow j$ for each j representing a position (F', Ψ') such that $F \xrightarrow{a} F'$ for $a \in K$
- If $\Psi = \nu Z_j.\, \Psi_j$, then $L(i) = V$ and there is the edge $i \rightarrow j'$ where j' represents (F, Z_j)
- If $\Psi = \mu Z_j.\, \Psi_j$, then $L(i) = R$ and there is the edge $i \rightarrow j'$ where j' represents (F, Z_j)
- If $\Psi = Z_j$ and $\nu Z_j.\, \Psi_j$ is in Sub(Φ), then $L(i) = V$ and there is the edge $i \rightarrow j'$ where j' represents (F, Ψ_j)
- If $\Psi = Z_j$ and $\mu Z_j.\, \Psi_j$ is in Sub(Φ), then $L(i) = R$ and there is the edge $i \rightarrow j'$ where j' represents (F, Ψ_j)

6. Finally, we tidy up and remove any positions not reachable from the initial position (E, Φ).

Consider any play of the resulting parity game $\mathsf{G_V}[E, \Phi]$. The least vertex i that occurs infinitely often must represent one of the following positions.

1. (F, Z) and Z is free in Φ
2. $(F, \mathtt{tt})$
3. $(F, \mathtt{ff})$
4. $(F, [K]\Psi)$ and $\{F' : F \xrightarrow{a} F' \text{ and } a \in K\} = \emptyset$
5. $(F, \langle K\rangle\Psi)$ and $\{F' : F \xrightarrow{a} F' \text{ and } a \in K\} = \emptyset$
6. (F, Z_j)

In all but the last case, there is the cycle $i \rightarrow i$, and the winner is the same as in the property checking game. In the parity game, these positions turn into loops to make playing perpetual. Otherwise, position i represents (F, Z_j). Consider any other j' occurring infinitely often and that represents position (F', Z_l). Because i is the least vertex occurring infinitely often, position (F', Z_l) appears later than (F, Z_j) in the position ordering. Therefore, either $Z_j = Z_l$ or the fixed point formula identified by Z_l is strictly smaller than that identified by Z_j; in which case, Z_j subsumes Z_l. If $\nu Z_j.\, \Psi_j$ is in Sub(Φ), then player V is the winner, and

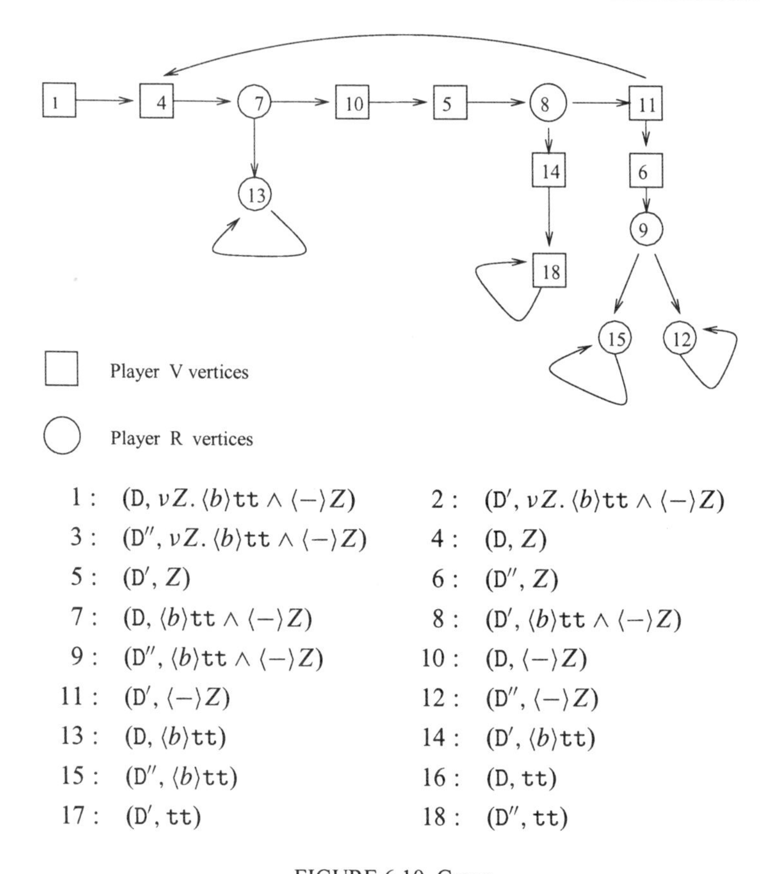

FIGURE 6.10. Game

instead if $\mu Z_j.\,\Psi_j$ is in Sub(Φ), then player R wins. This agrees with the property checking game.

The notions of a (history-free) strategy and of a winning strategy for the parity game are very similar to the property checking game. For player P, a history-free strategy is a set of rules of the form "at i choose j" where $L(i) = P$ and there is an edge $i \rightarrow j$. A strategy is winning for player P if she wins every play in which that strategy is employed. The following proposition is deducible from the analysis and observations above.

Proposition 1 *Player P has a winning strategy for $G_V(E, \Phi)$ iff Player P has a winning strategy for $G_V[E, \Phi]$.*

Example 1 The game G[D, $\nu Z.\,\langle b\rangle$tt $\wedge \langle -\rangle Z$], where D is depicted in Figure 6.3 is given in Figure 6.10. Vertices (such as 2 and 3) not reachable from vertex 1 in the game are excluded. The representation of positions from the model checking game are also presented.

Exercises

1. Define the parity games G[E, Φ] when E and Φ are the following.
   a. D and Ψ from Section 6.2.
   b. Ven and $\nu X. \langle - \rangle X$
   c. Crossing and $\mu X. [-] X$
2. Prove Proposition 1.
3. Prove that there is a converse to Proposition 1. Any parity game can be transformed into a property checking game whose size is polynomially bounded by the parity game. (See Mader for an explicit construction [39].)
4. Define infinite state parity games so that Proposition 1 also holds for infinite state processes E. (Hint: use a finite set of indexed colours.)
5. Boolean fixed point logic is defined as follows.

$$\Phi \ ::= \ Z \mid \mathtt{tt} \mid \mathtt{ff} \mid \Phi_1 \wedge \Phi_2 \mid \Phi_1 \vee \Phi_2 \mid \nu Z. \Phi \mid \mu Z. \Phi$$

   Formulas are interpreted over the two element set $\emptyset$ (false) and $\{1\}$ (true). A closed formula is therefore either true or false. For instance, $\mu Z. Z$ is false because $\emptyset$ is the least solution to the equation $Z = Z$.
   a. Show that a parity game can be directly translated into a closed formula of boolean fixed point logic in such a way that player V wins the game iff the translated formula is true.
   b. Show the converse, that any closed boolean formula can be translated into a game in such a way that the formula is true iff player V wins the game.
6. A simple stochastic game, SSG, is a graph game whose vertices are labelled R, V or A (average), and where there are two special vertices R-sink and V-sink (which have no outgoing edges). Each R and V vertex (other than the sinks) has at least one outgoing edge, and each A vertex has exactly two outgoing edges. At an average vertex during a game play a coin is tossed to determine which of the two edges is traversed, each having probability $\frac{1}{2}$. A game play ends when a sink vertex is reached: player V wins if it is the V-sink, and player R otherwise (which includes the eventuality of the game continuing forever). The decision question is whether the probability that player V wins is greater than $\frac{1}{2}$. It is not known whether this problem can be solved in polynomial time; see Condon [15].
   Show how to (polynomially) transform a parity game into an SSG in such a way that the same player wins both games. (Hint: add the two sink vertices, and an average vertex $i1$ for each vertex i for which there is an edge $j \longrightarrow i$ with $j \geq i$. Each such edge $j \longrightarrow i$ when $j \geq i$ is removed, and the edge $j \longrightarrow i1$ is added. Two new edges are added for each A vertex $i1$: first an edge to i, and second an edge to R-sink, if i is labelled R, or to V-sink if it is labelled V. The difficult question is: what probabilities these edges should have?)

6.6 Deciding parity games

The model checking question "is it the case that $E \models_V \Phi$?" is equivalent to the decision question "which player has a winning strategy for the parity game $G_V[E, \Phi]$?" In this section, we develop a method for deciding parity games and for exhibiting a winning strategy. The technique uses subgames of a parity game.

If $G = (N, \rightarrow, L)$ is a parity game and i is a vertex in N, then $G(i)$ is the game G, except that the starting vertex is i. The game G itself is $G(j)$, where j is the smallest vertex. If $X \subseteq N$, then $G - X$ is the result of removing all vertices in X from G, all edges from vertices in X, and all edges into vertices in X. It is therefore the subgraph $G' = (N', \rightarrow', L')$ whose vertices $N' = N - X$, and whose edge relation $\rightarrow'$ is $\rightarrow \bigcap (N' \times N')$, and further whose labelling $L'(j) = L(j)$ for each $j \in N'$. The subgraph G' may not be a game because it may not satisfy the requirement that every vertex $i \in N'$ have an edge $i \rightarrow' j$. If it does obey this condition, then G' is said to be a *subgame* of G. A degenerate case is when N' is the emptyset. Now we shall consider appropriate subsets X so that $G - X$ is guaranteed to be a subgame.

A useful notion is a set of vertices of a parity game for which a player P can force play to enter a subset X of vertices, written as $\text{Force}_P(X)$. This idea of a Force set is used in Section 3.2 in the case of bisimulation games. It was used by McNaughton and others [41, 58] in the case of more general games. A force set is defined iteratively as follows.

Definition 1 Let $G = (N, \rightarrow, L)$ be a parity game and $X \subseteq N$,

$$\text{Force}^0_P(X) = X \text{ for } P \in \{R, V\}$$

$$\begin{aligned}\text{Force}^{i+1}_R(X) &= \text{Force}^i_R(X)\\ &\cup \{j : L(j) = R \text{ and } \exists k \in \text{Force}^i_R(X).\, j \rightarrow k\}\\ &\cup \{j : L(j) = V \text{ and } \forall k.\text{ if } j \rightarrow k \text{ then } k \in \text{Force}^i_R(X)\}\end{aligned}$$

$$\begin{aligned}\text{Force}^{i+1}_V(X) &= \text{Force}^i_V(X)\\ &\cup \{j : L(j) = V \text{ and } \exists k \in \text{Force}^i_V(X).\, j \rightarrow k\}\\ &\cup \{j : L(j) = R \text{ and } \forall k.\text{ if } j \rightarrow k \text{ then } k \in \text{Force}^i_V(X)\}\end{aligned}$$

$$\text{Force}_P(X) = \bigcup\{\text{Force}^i_P(X) : i \geq 0\} \text{ for } P \in \{R, V\}.$$

If vertex $j \in \text{Force}_P(X)$ and the current position is j, then player P can force play from j into X irrespective of whatever moves her opponent makes. Vertex j itself need not belong to player P. The *rank* of such a vertex j is the least index i such that $j \in \text{Force}^i_P(X)$. The rank is an upper bound on the total number of moves it takes for player P to force play into X. There is an associated strategy for player P.

For every vertex $i \in \text{Force}_P(X)$ belonging to P, either $i \in X$, or there is an edge $i \to k$ and $k \in \text{Force}_P(X)$, so the strategy for P is to choose a k with the least rank.

The definition of a force set provides a method for computing it. As i increases, we calculate $\text{Force}^i_P(X)$ until it is the same set as $\text{Force}^{i-1}_P(X)$. Clearly, this must hold when $i \leq (|N| - |X|) + 1$.

Example 1 Consider the following force set, where the vertices are from Figure 6.10.

$$
\begin{aligned}
\text{Force}^0_V(\{12, 15\}) &= \{12, 15\} \\
\text{Force}^1_V(\{12, 15\}) &= \{12, 15\} \cup \{9\} \\
\text{Force}^2_V(\{12, 15\}) &= \{9, 12, 15\} \cup \{6\} \\
\text{Force}^3_V(\{12, 15\}) &= \{6, 9, 12, 15\} \cup \{11\} \\
\text{Force}^4_V(\{12, 15\}) &= \{6, 9, 11, 12, 15\} \cup \emptyset
\end{aligned}
$$

So, $\text{Force}_V(\{12, 15\}) = \{6, 9, 11, 12, 15\}$, and 11 has rank 3. The different set $\text{Force}_R(\{12, 15\}) = \{6, 9, 12, 15\}$.

The following result shows that removing a force set from a game leaves a subgame.

Proposition 1 *If* G *is a game and X is a subset of vertices, then the subgraph* $\mathsf{G} - \text{Force}_P(X)$ *is a subgame.*

Proof. Assume that $\mathsf{G} = (N, \to, L)$ is a game and $X \subseteq N$, and that P is a player. Consider the structure $\mathsf{G}' = \mathsf{G} - \text{Force}_P(X)$. G' fails to be a subgame if there is a vertex j in N' such that there is no edge $j \to' k$. Consider any such vertex j. Clearly, each k such that $j \to k$ belongs to $\text{Force}_P(X)$, so there is a least index i such that $k \in \text{Force}^i_P(X)$ for each such k. But then $j \in \text{Force}^{i+1}_P(X)$, and therefore $j \in \text{Force}_P(X)$, which contradicts $j \in N'$. □

We wish to provide a decision procedure for parity games that not only computes the winner but also a winning strategy. Such a procedure can be extracted from the proof of the following theorem.

Theorem 1 *For any parity game* G *and vertex i, one of the players has a history-free winning strategy for* $\mathsf{G}(i)$.

Proof. Let $\mathsf{G} = (N, \to, L)$ be a parity game. The proof is by induction on $|N|$. The base case is when $|N| = 0$ and the result holds. For the inductive step, let $|N| > 0$ and assume that k is the least vertex in N. Let X be the set $\text{Force}_{L(k)}(\{k\})$.

If $X = N$, then player $L(k)$ has a history-free winning strategy for $\mathsf{G}(i)$ for each $i \in N$, by forcing play infinitely often through vertex k. More precisely, the strategy consists of the rules "at j choose $j1$" when $L(j) = L(k)$ and where $j1$ has a least rank in $\text{Force}_{L(k)}(\{k\})$ among the set $\{j' : j \to j'\}$: this strategy ensures that k will be traversed infinitely often in every play.

Otherwise, $X \neq N$. By Proposition 1, $\mathsf{G}' = \mathsf{G} - X$ is a subgame. The size of N' is strictly smaller than the size of N. Therefore, by the induction hypothesis, for each $j \in N'$, player P_j has a history-free winning strategy σ_j' for the game $\mathsf{G}'(j)$. Partition these vertices into W_R', the vertices won by player R, and W_V', the vertices won by player V, as in Figure 6.11. Notice that, with respect to G', the set $W_P' = \text{Force}_P(W_P')$ when P is R or V. The proof now consists of examining two subcases, depending on the set $Y = \{j : k \rightarrow j\}$.

Case 1 $Y \cap (W_{L(k)}' \cup X) \neq \emptyset$. There is an edge $k \rightarrow j1$ and vertex $j1 \in X$ or $j1 \in W_{L(k)}'$, as in Figure 6.11. Player $L(k)$ has a history-free winning strategy for the subgame $\mathsf{G}(i)$ for each $i \in X \cup W_{L(k)}'$. Let π be the substrategy for $L(k)$ that forces play from any vertex in X to vertex k, as described earlier. Add to π the rule "at k choose $j1$", and let π' be the resulting strategy. If $j1 \in X$, then π' is a history-free winning strategy for $\mathsf{G}(i)$ for each $i \in X$, and $\sigma_i' \cup \pi'$ is a history-free winning strategy for $\mathsf{G}(i)$ for any $i \in W_{L(k)}'$: the opponent may move from $W_{L(k)}'$ into X. If $j1 \in W_{L(k)}'$, then the strategy $\pi' \cup \sigma_{j1}'$ is winning for $\mathsf{G}(i)$ for $i \in X \cup \{j1\}$, and $\sigma_i' \cup \pi' \cup \sigma_{j1}'$ is a winning strategy for $\mathsf{G}(i)$ for i in $W_{L(k)}'$. Again, the opponent may move from $W_{L(k)}'$ into X. The opponent O of $L(k)$ has the history-free winning strategy σ_i' for each game $\mathsf{G}(i)$ when $i \in W_O'$.

Case 2 $Y \cap (W_{L(k)}' \cup X) = \emptyset$. This means that, for every $j1$ such that $k \rightarrow j1$, the opponent O of $L(k)$ has a history-free winning strategy for $\mathsf{G}'(j1)$. Let $Z = \text{Force}_O(W_O')$ with respect to the full game G (see the shaded picture in Figure 6.11): notice that Z contains the vertex k and possibly vertices from $W_{L(k)}'$. For each $i \in Z$ player O has a history-free winning strategy for $\mathsf{G}(i)$: the strategy consists of forcing play into W_O' and then using the winning strategies determined from G', much like we described. Let σ_i be the history-free strategy for any $i \in Z$. If $Z = N$, then σ_i is the strategy for $\mathsf{G}(i)$. Otherwise, consider the subgame $\mathsf{G}'' = \mathsf{G} - Z$ (the unshaded part of the game in Figure 6.11). By the induction hypothesis, for each j in G'' player P_j has a history-free winning strategy σ_j'' for $\mathsf{G}''(j)$. If $P_j = L(k)$, then σ_j'' is a history-free winning strategy for $L(k)$ for the game $\mathsf{G}(j)$. Otherwise, $P_j = O$, and player O has a history-free winning strategy for $\mathsf{G}(j)$. Player O uses the partial strategy σ_j'' until (if at all) player $L(k)$ plays into the set Z, in which case player O uses the approriate winning strategy, which keeps the play in Z. We leave the details to the reader. □

The proof of Theorem 1 contains a recursive algorithm for model checking, which also computes winning strategies. Notice that a winning strategy is linear in the size of the game. The decision question is, given a parity game $\mathsf{G} = (N, \rightarrow, L)$, determine for each vertex i what player wins $\mathsf{G}(i)$, and what the winning strategy is. Below is a summary of the algorithm that leaves out the computation of winning strategies, which the reader can add. It computes the sets W_R and W_V, which are the vertices that player R wins and the vertices that player V wins.

1. Let k be the least vertex in N, let $X = \text{Force}_{L(k)}(\{k\})$ and let O be the opponent of $L(k)$
2. If $X = N$, then return $W_{L(k)} = N$ and $W_O = \emptyset$

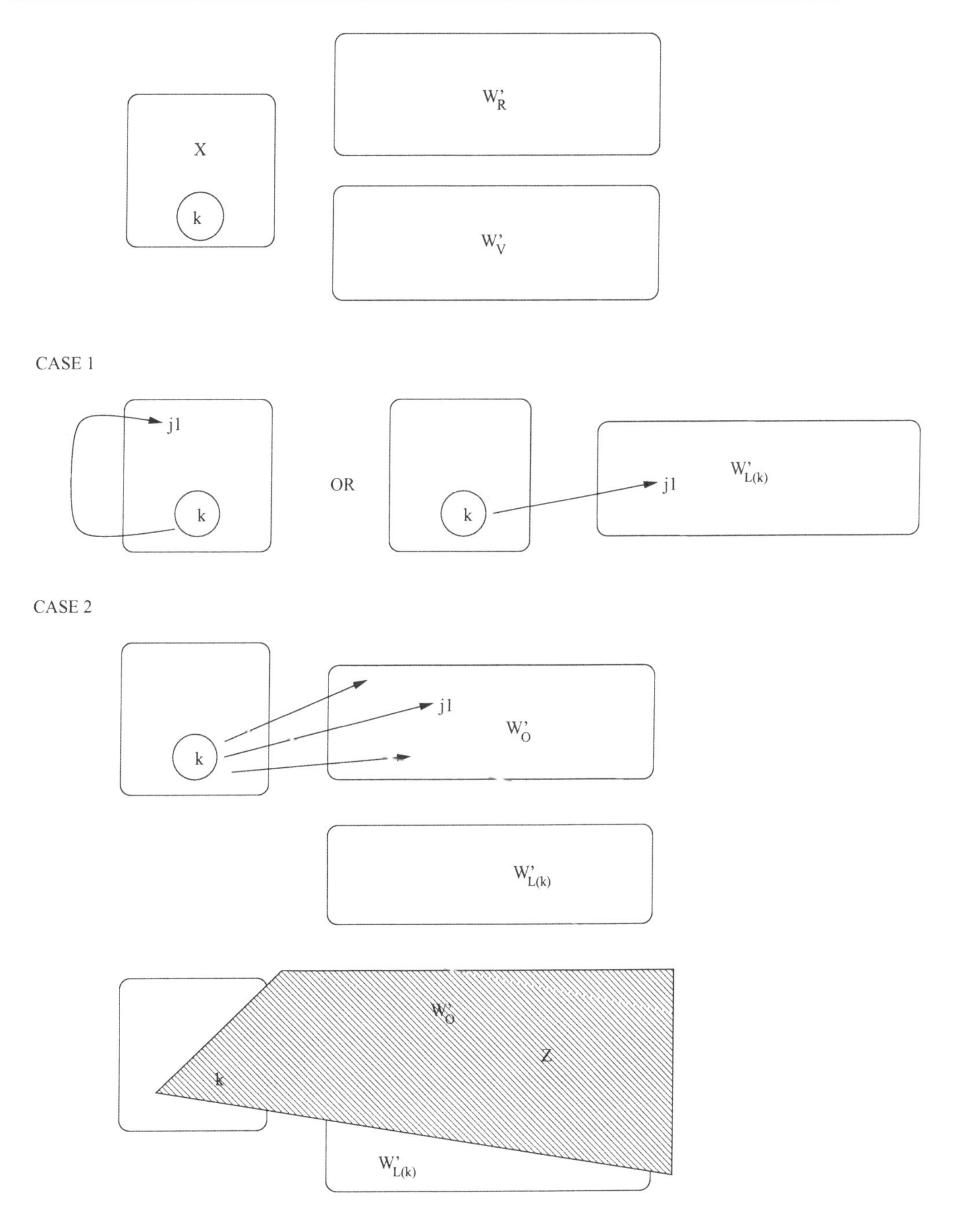

FIGURE 6.11. Cases in Theorem 1

3. Else solve the subgame $\mathsf{G} - X$, and let W'_R and W'_V be the winning vertices for R and V in the subgame. Let $Y = \{j : k \rightarrow j\}$
   a. If $Y \cap (W'_{L(k)} \cup X) \neq \emptyset$, then return $W_{L(k)} = W'_{L(k)} \cup X$ and also $W_O = W'_O$
   b. Else let $Z = \text{Force}_O(W_O)$. Solve the subgame $\mathsf{G} - Z$, and let W''_R and W''_V be the winning vertices for R and V in the subgame. Return $W_O = Z \cup W''_O$ and $W_{L(k)} = W''_{L(k)}$

The algorithm in Theorem 1 is exponential in the size of N because it may twice call the solve procedure on games of smaller size at stage 3, and then again at stage 3b. It is an open question whether there is a polynomial time algorithm for this problem. We shall now look at subcases where model checking can be done more quickly.

The first case is for alternation free formulas, the sublogic μMA that contains μMI and also CTL. Recall that if Φ is a CTL formula, then the game $\mathsf{G}(E, \Phi)$ has the property that, in any infinite length run, there is just one until formula, or the negation of an until formula, occurring infinitely often. More generally (see Proposition 2 of Section 6.4) if Φ belongs to μMA, then the game $\mathsf{G}_\mathsf{V}(E, \Phi)$ has the property that, in any infinite length run all the variables that occur infinitely often are of the same kind. This means that the resulting parity game $\mathsf{G}_\mathsf{V}[E, \Phi]$ has a pleasant structure. Its vertices can be partitioned into a family of subsets $N_1, \ldots, N_n$ such that the following properties hold.

1. $N_i \cap N_j = \emptyset$ when $i \neq j$
2. if any play is at a vertex in N_j, then the play cannot return to a vertex in N_i when $i < j$
3. for each i, all plays that remain in N_i forever are always won by the same player

For such a parity game there is a straightforward iterative linear time algorithm that finds the winner as follows. Start with the subgame whose vertices are N_n. This must be a subgame; for there are no edges from N_n to N_j, $j < n$. One of the players wins every vertex of N_n. Suppose it is player P. Let $X_n = \text{Force}_P(N_n)$. Player P wins every vertex in X_n. Let $\mathsf{G}' = \mathsf{G} - X_n$ and assume its vertex set is N'. This is a subgame. Consider the non-empty sets in the partition $N_1 \cap N', \ldots, N_{n-1} \cap N'$. These sets will inherit the properties above. Consider the final such set. One of the players wins every vertex in this set, and so on.

The second case we consider is when one of the players never has a choice of move. The game obeys the following condition in case of one of the players P.

$$\text{for any vertex } i, \text{ if } L(i) = P \text{ and } i \rightarrow j \text{ and } i \rightarrow k, \text{ then } j = k$$

Games display a subgame invariance property for the opponent O. Let Q be one of the players, and let $\mathsf{G}' = \mathsf{G} - \text{Force}_\mathsf{Q}(X)$. If player O wins vertex $i \in \mathsf{G}'$, then she also wins $i \in \mathsf{G}$ because player P cannot escape from G' into $\text{Force}_\mathsf{Q}(X)$.

Therefore, we can use Theorem 1 to provide a polynomial time algorithm for this case.

Exercises

1. Consider vertices in the parity game of Figure 6.10. Work out the following force sets.
   a. $\mathrm{Force}_{\mathrm{V}}(\{18\})$
   b. $\mathrm{Force}_{\mathrm{R}}(\{18\})$
   c. $\mathrm{Force}_{\mathrm{V}}(\{13, 15, 12\})$
   d. $\mathrm{Force}_{\mathrm{R}}(\{13, 15, 12\})$
2. Use the algorithm from Theorem 1 to decide who wins each vertex, and what the winning strategy is, for the game in Figure 6.10.
3. Emerson, Jutla and Sistla show that the model checking decision problem belongs to NP ∩ co-NP because deciding who wins a parity game belongs to NP, and model checking is closed under complement [20]. Give a proof that deciding who wins a parity game belongs to NP.
4. Prove that, if $\Phi \in \mu\mathrm{MA}$, then the parity game $\mathrm{G_V}[E, \Phi]$ has the pleasant structure decribed. That is, its vertices can be partitioned into a family of subsets $N_1, \ldots, N_n$ such that the following properties hold.
   a. $N_i \cap N_j = \emptyset$ when $i \neq j$
   b. if any play is at a vertex in N_j, then the play cannot return to a vertex in N_i when $i < j$
   c. for each i, all plays that remain in N_i forever are always won by the same player
5. Assume the parity game G obeys the following condition for player P whose opponent is O:

   for any vertex i, if $L(i) = P$ and $i \to j$ and $i \to k$, then $j = k$.

   Prove that, if Q is one of the players, and $\mathrm{G}' = \mathrm{G} - \mathrm{Force}_{\mathrm{Q}}(X)$ and player O wins vertex $i \in \mathrm{G}'$, then she also wins $i \in \mathrm{G}$. Use this fact to provide a quick algorithm for deciding parity games obeying this condition.

7

Exposing Structure

In the previous chapter, an alternative characterization of when a process has a property was presented in terms of games. Simple property checking games were presented where the verifier has a winning strategy for a game if, and only if, the process satisfies the formula. When the state space of a finite state process is large, a proof using games may become unwieldy. Moreover, we should like to be able to cope with processes that have infinite state spaces. Plays of games involving such processes may have infinitely many different positions. The question then arises as to when there can be a finitely presented strategy, as a summary of the successful strategy for a player.

There is another reason for examining more general ideas. Often we are interested in showing properties of schematic or parameterised processes. Processes involving value passing constitute one such family. Another kind of example is illustrated by the scheduler $\mathtt{Sched}_n$ of Section 1.4, which is parametric on the size of the cycle n. The techniques of the previous chapter allow us to prove that, for instance, $\mathtt{Sched}_{32}$ is free from deadlock. However, they do not allow us to directly show that $\mathtt{Sched}_n$ for any $n > 1$ is deadlock free.

7.1 Infinite state systems

Process definitions may involve explicit parameterisation. Examples include the counters $\mathtt{Ct}_i$ and registers $\mathtt{Reg}_i$ of Section 1.1, and the slot machines $\mathtt{SM}_n$ of Section 1.2. Each instantiation of these processes, when i and n are set to a particular value, is itself infinite state and contains the other family members within its transition graph. However, the parameterisation is very useful because it reveals straightforward structural similarities within these families of processes.

Another class of processes that is infinite state owes entirely to the presence of data values. This family includes the copier Cop and the processes $\mathtt{T}(i)$ of Section 1.1 and the protocol Protocol of Section 1.2. However, there are different degrees of involvement of data within these processes, depending on the extent to which data determines future behaviour. At one extreme are examples such as Cop and Protocol, which pass data items through the system oblivious to their values. Different authors have identified classes of processes that are in this sense data independent. At the other extreme are systems such as $\mathtt{T}(i)$, where future behaviour strongly depends on the value i. In between are systems such as the registers $\mathtt{Reg}_i$, where particular values are essential to change of state.

A third class of processes is infinite state independently of parameterisation and data values. An instance is the counter Count of Section 1.5, which evolves its structure as it performs actions. In certain cases, processes that are infinite state, in that they determine an infinite state transition graph, are in fact bisimulation equivalent to a finite state process. A simple example is that C and C′ are bisimilar, where these processes are as follows.

$$\begin{aligned} \mathtt{C} &\stackrel{\text{def}}{=} a.\mathtt{C} \mid b.\mathtt{C} \\ \mathtt{C}' &\stackrel{\text{def}}{=} a.\mathtt{C}' + b.\mathtt{C}' \end{aligned}$$

Although the structure of C evolves through behaviour, $\mathtt{C} \stackrel{a}{\longrightarrow} \mathtt{C} \mid b.\mathtt{C}$ for example, and therefore the transition graph for C is infinite state, each state of the graph is bisimulation equivalent to the starting state, and hence to $\mathtt{C}'$. A more general interesting subclass of processes consists of those that can be infinite state, but for which bisimulation equivalence is decidable. Two examples are context free processes and basic parallel processes[1]; see Hirshfeld and Moller for a survey of results [30].

A final class of systems is also parameterised. However, for each instance of the parameter the system is finite state. Two paradigm examples are the buffer $\mathtt{Buff}_n$ and the scheduler $\mathtt{Sched}_n$, both from Section 1.4. Although the techniques for verification of temporal properties apply to intances they do not apply to the general families. In such cases, we should like to prove properties generally, to show for instance that $\mathtt{Sched}_n$ is free from deadlock for each $n > 1$. The proof of

[1] The processes C and C′ are basic parallel processes.

this requires us to expose structure that is common to the whole family of processes. In this case, of freedom from deadlock, the property itself is not parameterised. Much more complex is the situation wherein it is. An example is again the case of the scheduler, that $\texttt{Sched}_n$ has the property Cycle($a_1 \ldots a_n$) for every $n > 1$ (where the parameterised property is defined in Section 5.7).

Exercises

1. **a.** Provide a definition for when a value passing process is data independent.
   b. Can you generalise your definition for when a value passing process only depends on finitely many different values (and so is "almost" data independent).

   Compare your definitions with those of Jonsson and Parrow [32].
2. Show that $\texttt{C} \sim \texttt{C}'$, where these processes are given above.
3. Prove the following for all $n > 1$.
   a. $\texttt{Sched}_n \models \nu Z.\,\langle - \rangle \texttt{tt} \wedge [-]Z$
   b. $\texttt{Sched}_n \models \text{Cycle}(a_1 \ldots a_n)$
4. Assume $\texttt{Cy}_n \stackrel{\text{def}}{=} a_1.\ldots a_n \texttt{Cy}_n$, for any $n > 1$.
   a. Prove that $\texttt{Sched}_n \backslash\backslash \{b_1, \ldots, b_n\} \approx \texttt{Cy}_n$, for all $n > 1$
   b. Now provide a definition of a parameterised bisimulation relation covering the example in part (a)

7.2 Generalising satisfaction

The satisfaction relation between an individual process and a property is not sufficiently general to capture properties of parameterised processes. Although it allows us to express, for instance, $\texttt{Sched}_4 \models \nu Z.\,\langle - \rangle \texttt{tt} \wedge [-]Z$, that $\texttt{Sched}_4$ is deadlock free, it does not allow the more general claim that $\texttt{Sched}_n$ for all $n > 1$ has this property. A straightforward remedy is to generalise the satisfaction relation so that it holds between a family of processes and a formula. We use the same relation $\models_{\mathsf{V}}$ for this extension.

$$\mathsf{E} \models_{\mathsf{V}} \Phi \text{ iff } E \models_{\mathsf{V}} \Phi \text{ for all } E \in \mathsf{E}$$

We write $\mathsf{E} \models \Phi$ when the formula contains no free variables. The more general claim that the schedulers are deadlock free is therefore represented as follows.

$$\{\texttt{Sched}_n : n > 1\} \models \nu Z.\,\langle - \rangle \texttt{tt} \wedge [-]Z$$

This generalization does not cover all possibilities discussed in the previous section. In particular, we leave as an exercise the situation wherein the property is also parameterised.

Example 1 The family of counters of Figure 1.4, $\{\texttt{Ct}_i : i \geq 0\}$, has the property $[\texttt{up}]([\texttt{round}]\texttt{ff} \wedge [\texttt{up}]\langle\texttt{down}\rangle\langle\texttt{down}\rangle\texttt{tt})$. The following "proof" uses properties of the generalised satisfaction relation, which are formalised precisely below.

$$
\begin{array}{ll}
 & \{\texttt{Ct}_i : i \geq 0\} \models [\texttt{up}]([\texttt{round}]\texttt{ff} \wedge [\texttt{up}]\langle\texttt{down}\rangle\langle\texttt{down}\rangle\texttt{tt}) \\
\text{iff} & \{\texttt{Ct}_i : i \geq 1\} \models [\texttt{round}]\texttt{ff} \wedge [\texttt{up}]\langle\texttt{down}\rangle\langle\texttt{down}\rangle\texttt{tt} \\
\text{iff} & \{\texttt{Ct}_i : i \geq 1\} \models [\texttt{round}]\texttt{ff} \text{ and } \{\texttt{Ct}_i : i \geq 1\} \models [\texttt{up}]\langle\texttt{down}\rangle\langle\texttt{down}\rangle\texttt{tt} \\
\text{iff} & \{\texttt{Ct}_i : i \geq 1\} \models [\texttt{up}]\langle\texttt{down}\rangle\langle\texttt{down}\rangle\texttt{tt} \\
\text{iff} & \{\texttt{Ct}_i : i \geq 2\} \models \langle\texttt{down}\rangle\langle\texttt{down}\rangle\texttt{tt} \\
\text{iff} & \{\texttt{Ct}_i : i \geq 1\} \models \langle\texttt{down}\rangle\texttt{tt} \\
\text{iff} & \{\texttt{Ct}_i : i \geq 0\} \models \texttt{tt}
\end{array}
$$

Arguably, this is a more direct proof than to appeal to induction on process indices. The reader is invited to prove it inductively, and expose the required induction hypothesis.

Example 1 uses various features of the satisfaction relation between sets of processes and formulas. The case when a formula is a conjunction of two subformulas is the most straightforward, $\mathsf{E} \models_{\mathsf{V}} \Phi \wedge \Psi$ iff $\mathsf{E} \models_{\mathsf{V}} \Phi$ and also $\mathsf{E} \models_{\mathsf{V}} \Psi$. Disjunction is more complicated. If $\mathsf{E} \models_{\mathsf{V}} \Phi \vee \Psi$, then this does not imply that $\mathsf{E} \models_{\mathsf{V}} \Phi$ or $\mathsf{E} \models_{\mathsf{V}} \Psi$. A simple illustration is that, although $\{0, a.0\} \models \langle a\rangle\texttt{tt} \vee [a]\texttt{ff}$, it is neither the case that $\{0, a.0\} \models \langle a\rangle\texttt{tt}$, nor is it the case that $\{0, a.0\} \models [a]\texttt{ff}$. In general, if $\mathsf{E} \models_{\mathsf{V}} \Phi \vee \Psi$, then some processes in E may have the property Φ and the rest have the property Ψ. Therefore, E can be spit into two subsets E_1 and E_2 in such a way that $\mathsf{E}_1 \models_{\mathsf{V}} \Phi$ and $\mathsf{E}_2 \models_{\mathsf{V}} \Psi$. One of these sets could be empty[2].

To understand the situation when a set of processes satisfies a modal formula, a little notation is introduced. If K is a set of actions, and E a set of processes, then $K(\mathsf{E})$ is the following set.

$$K(\mathsf{E}) \stackrel{\text{def}}{=} \{F : E \stackrel{a}{\longrightarrow} F \text{ for some } E \in \mathsf{E} \text{ and } a \in K\}$$

Therefore, $K(\mathsf{E})$ is the set of processes reachable by K transitions from members of E. For instance, $\{\texttt{up}\}(\{\texttt{Ct}_i : i \geq 0\})$ in example 1 is the set $\{\texttt{Ct}_i : i \geq 1\}$. Clearly, $\mathsf{E} \models_{\mathsf{V}} [K]\Phi$ iff $K(\mathsf{E}) \models_{\mathsf{V}} \Phi$. This principle is used in example 1. One case[3] is $\{\texttt{Ct}_i : i \geq 1\} \models [\texttt{up}]\langle\texttt{down}\rangle\langle\texttt{down}\rangle\texttt{tt}$ iff

$$\{\texttt{up}\}(\{\texttt{Ct}_i : i \geq 1\} \models \langle\texttt{down}\rangle\langle\texttt{down}\rangle\texttt{tt}.$$

A function $f : \mathsf{E} \rightarrow K(\mathsf{E})$ that maps processes in E into processes in $K(\mathsf{E})$ is a "choice function" if, for each $E \in \mathsf{E}$, there is an $a \in K$ such that $E \stackrel{a}{\longrightarrow} f(E)$. If $f : \mathsf{E} \rightarrow K(\mathsf{E})$ is a choice function, then $f(\mathsf{E})$ is the set of processes

[2] By definition $\emptyset \models_{\mathsf{V}} \Phi$ for any Φ.

[3] Another use of the principle in example 1 is that $\{\texttt{Ct}_i : i \geq 1\} \models [\texttt{round}]\texttt{ff}$ because $\{\texttt{round}\}(\{\texttt{Ct}_i : i \geq 1\}) = \emptyset$ and $\emptyset \models \texttt{ff}$.

$\mathsf{E} \models_{\mathsf{V}} Z$	iff	$\mathsf{E} \subseteq \mathsf{V}(Z)$
$\mathsf{E} \models_{\mathsf{V}} \Phi \wedge \Psi$	iff	$\mathsf{E} \models_{\mathsf{V}} \Phi$ and $\mathsf{E} \models_{\mathsf{V}} \Psi$
$\mathsf{E} \models_{\mathsf{V}} \Phi \vee \Psi$	iff	$\mathsf{E} = \mathsf{E}_1 \cup \mathsf{E}_2$ and $\mathsf{E}_1 \models_{\mathsf{V}} \Phi$ and $\mathsf{E}_2 \models_{\mathsf{V}} \Psi$
$\mathsf{E} \models_{\mathsf{V}} [K]\Phi$	iff	$K(\mathsf{E}) \models_{\mathsf{V}} \Phi$
$\mathsf{E} \models_{\mathsf{V}} \langle K \rangle\Phi$	iff	there is a choice function $f : \mathsf{E} \to K(\mathsf{E})$ and $f(\mathsf{E}) \models_{\mathsf{V}} \Phi$
$\mathsf{E} \models_{\mathsf{V}} \sigma Z.\Phi$	iff	$\mathsf{E} \models_{\mathsf{V}} \Phi\{\sigma Z.\Phi/Z\}$

FIGURE 7.1. Semantics for $\mathsf{E} \models_{\mathsf{V}} \Phi$

$\{f(E) : E \in \mathsf{E}\}$. Example 1 uses the choice function $f : \{\mathtt{Ct}_i \,;\, i \geq 2\} \to \{\mathtt{down}\}(\{\mathtt{Ct}_i \,;\, i \geq 2\})$ where $f(\mathtt{Ct}_{i+1}) = \mathtt{Ct}_i$. Choice functions are appealed to when understanding satisfaction in the case of $\langle K \rangle$ modal formulas, $\mathsf{E} \models_{\mathsf{V}} \langle K \rangle\Phi$ iff there is a choice function $f : \mathsf{E} \to K(\mathsf{E})$ such that $f(\mathsf{E}) \models \Phi$. Consequently, in example 1 $\{\mathtt{Ct}_i \,;\, i \geq 1\} \models \langle\mathtt{down}\rangle\mathtt{tt}$ iff $\{\mathtt{Ct}_i \,;\, i \geq 0\} \models \mathtt{tt}$.

The final cases to examine are the fixed points. We shall make use of the principles developed in the previous chapter using games. Notice however, that the fixed point unfolding principle, Proposition 2 of Section 5.1, holds for the generalised satisfaction relation. In Figure 7.1 we summarise the conditions for satisfaction between a family of processes and a formula.

Exercises

1. Prove example 1 using the principles of Figure 7.1. Also provide a proof that $\mathtt{Ct}_i \models [\mathtt{up}]([\mathtt{round}]\mathtt{ff} \wedge [\mathtt{up}]\langle\mathtt{down}\rangle\langle\mathtt{down}\rangle\mathtt{tt})$ for all $i \geq 0$ using induction on i. What induction hypothesis do you appeal to?
2. Show the following
   a. $\{\mathtt{Ct}_i \;:\, i \geq 0\} \not\models [\mathtt{up}]\langle\mathtt{down}\rangle\langle\mathtt{down}\rangle\mathtt{tt}$
   b. $\{\mathrm{T}(i) \,:\, i \leq 10\} \models \mu Y.\, \langle -\rangle\mathtt{tt} \wedge [-\overline{\mathtt{out}}(1)]Y$
   c. $\{\mathtt{Sched}_n \,:\, n > 1\} \models \mathrm{Cycle}(a_1 a_2)$

   where $\mathrm{T}(i)$ is defined in Section 1.1.
3. Can you show the following?

   $$\{\mathtt{Sem}^n \,:\, n \geq 1\} \models \nu Z.\, [\mathtt{get}](\mu Y.\, \langle -\rangle\mathtt{tt} \wedge [-\mathtt{put}]Y) \wedge [-]Z,$$

   where $\mathtt{Sem} \stackrel{\mathrm{def}}{=} \mathtt{get.put.Sem}$ and

   $$\mathtt{Sem}^1 \;=\; \mathtt{Sem}$$
   $$\mathtt{Sem}^{i+1} \;=\; \mathtt{Sem} \mid \mathtt{Sem}^i \quad i \geq 1.$$
4. Assume a copier shared by n users, where $n \geq 2$, given as a system Sys which is $(\mathtt{Cop} \mid \mathtt{User}_1 \mid \ldots \mid \mathtt{User_n})\backslash\{\mathtt{in}\}$ whose components are defined as

follows.

$$\texttt{Cop} \stackrel{\text{def}}{=} \texttt{in}(x).\overline{\texttt{out}}(x).\texttt{Cop}$$

$$\texttt{User}_i \stackrel{\text{def}}{=} \texttt{write}(x).\overline{\texttt{in}}(x).\texttt{User}_i$$

Demonstrate that this system has the property that, if v_1 and then v_2 are written, then the output of either of these values may be the next observable action.

5. Develop similar principles to those in Figure 7.1 that also allow parameterisation of formulas. Can you prove the following using these principles, for all $n > 1$?

$$\texttt{Sched}_n \models \text{Cycle}(a_1 \ldots a_n)$$

7.3 Tableaux I

In the previous chapter, games were developed for checking properties of individual processes. Our interest is now in checking properties of sets of processes. Instead of defining games to do this, we provide a tableau proof system for proving properties of sets of processes. A tableau proof system is goal directed, similar in spirit to the mechanism in Chapter 1 for deriving transitions. Its main ingredient is a finite family of proof rules that allow reduction of goals to subgoals. Tableaux have been used for property checking of both finite state and infinite state systems; see Bradfield, Cleaveland, Larsen and Walker [8, 10, 11, 13, 36, 55].

A goal has the form $\mathsf{E} \vdash_{\mathsf{V}} \Phi$ where E is a set of processes, Φ is a normal formula as defined in Section 5.2 and V is a valuation. The intention is to try to achieve the goal $\mathsf{E} \vdash_{\mathsf{V}} \Phi$, that is, to show that it is true, that $\mathsf{E} \models_{\mathsf{V}} \Phi$. The proof rules allow us to reduce goals to subgoals, and each rule has one of the following two forms:

$$\frac{\mathsf{E} \vdash_{\mathsf{V}} \Phi}{\mathsf{F} \vdash_{\mathsf{V}} \Psi}\ C \qquad\qquad \frac{\mathsf{E} \vdash_{\mathsf{V}} \Phi}{\mathsf{F}_1 \vdash_{\mathsf{V}} \Psi_1 \quad \mathsf{F}_2 \vdash_{\mathsf{V}} \Psi_2}\ C,$$

where C is a side condition. In both cases, the premise, $\mathsf{E} \vdash_{\mathsf{V}} \Phi$, is the goal that we are trying to achieve (that E has the property Φ relative to V), and the subgoals in the consequent are what the goal reduces to.

The tableau proof rules are presented in Figure 7.2. Other than Thin, each rule operates on the main logical connective of the formula in the goal. The logical rules for boolean and modal connectives follow the stipulations described in Figure 7.1. For instance, to establish the goal $\mathsf{E} \vdash_{\mathsf{V}} \Phi \wedge \Psi$, it is necessary to establish that E has the property Φ, and that E has the property Ψ, both relative to V. A fixed point formula is abbreviated by its bound variable, since we are dealing with normal formulas. The rule for a bound variable is just the fixed point unfolding rule. Thin is a structural rule, which allows the set of processes in a goal to be expanded.

The rules are backwards sound, which means that, if all the subgoals in the consequent of a rule are true, then so is the premise goal. The stipulations introduced

$$\wedge \quad \frac{E \vdash_V \Phi \wedge \Psi}{E \vdash_V \Phi \quad E \vdash_V \Psi}$$

$$\vee \quad \frac{E \vdash_V \Phi \vee \Psi}{E_1 \vdash_V \Phi \quad E_2 \vdash_V \Psi} \; E = E_1 \cup E_2$$

$$[K] \quad \frac{E \vdash_V [K]\Phi}{K(E) \vdash_V \Phi}$$

$$\langle K \rangle \quad \frac{E \vdash_V \langle K \rangle \Phi}{f(E) \vdash_V \Phi} \; f : E \rightarrow K(E) \text{ is a choice function}$$

$$\sigma Z. \quad \frac{E \vdash_V \sigma Z.\, \Phi}{E \vdash_V Z}$$

$$Z \quad \frac{E \vdash_V Z}{E \vdash_V \Phi} \; Z \text{ identifies the subformula } \sigma Z.\, \Phi$$

$$\text{Thin} \quad \frac{E \vdash_V \Phi}{F \vdash_V \Phi} \; E \subset F$$

FIGURE 7.2. Tableaux rules

in the previous section justify this for the boolean and modal operators, and in the case of the fixed point rules this follows from the fixed point unfolding property, Proposition 2 of Section 5.1. Backwards soundness of Thin is also clear because, if $F \models_V \Phi$ and $E \subset F$, then $E \models_V \Phi$.

The other ingredient of a tableau proof system is when a goal counts as a "terminal" goal, so that no rule is then applied to it. As we shall see, terminal goals are either "successful" or "unsuccessful." To show that all the processes in E have the property Φ relative to V, we try to achieve the goal $E \vdash_V \Phi$ by building a successful tableau whose root is this initial goal. A successful tableau is a finite proof tree whose root is the initial goal, and all whose leaves are successful terminal goals. Intermediate subgoals of the proof tree are determined by an application of one of the rules to the goal immediately above them.

The definition of when a goal in a tableau is terminal is underpinned by the game theoretic characterization of satisfaction. A tableau represents a family of games. Each process in the set of processes of a goal determines a play. The definition of when a goal $F \vdash_V \Psi$ in a proof tree is terminal is presented in Figure 7.3. Clearly, goals fulfilling 1 or 2 are correct. For instance, $F \not\models_V \langle K \rangle \Phi$ if there is an $F \in F$ and $F \not\models_V \langle K \rangle \Phi$. The justification for the success of condition 3 in the case of a successful terminal follows from considering any infinite length game play from a process in E with respect to the property Z that "cycles" through the terminal goal of the proof tree $F \vdash_V Z$. Because $F \subseteq E$, the play jumps back to the companion goal (the goal $E \models_V Z$). Such an infinite play must pass through the variable Z infinitely often, and because Z subsumes any other variable that also occurs infinitely often, and Z identifies a maximal fixed point formula, the play is therefore a win for the verifier. The justification for the failure of condition 3 in the case of an unsuccessful leaf is more involved, and will be properly accounted for in the next section when correctness of the proof method is shown. However, notice that if $E = F$, then any infinite length play from a process in E that repeatedly

Successful terminal $\mathsf{F} \vdash_{\mathsf{V}} \Psi$

1. $\Psi = \texttt{tt}$ or $\Psi = Z$ and Z is free in the initial formula and $\mathsf{F} \subseteq \mathsf{V}(Z)$
2. $\mathsf{F} = \emptyset$,
3. $\Psi = Z$ and there is a goal $\mathsf{E} \vdash_{\mathsf{V}} Z$ above $\mathsf{F} \vdash_{\mathsf{V}} Z$, such that there is a path as follows

$$\begin{array}{ll} \mathsf{E} \vdash_{\mathsf{V}} Z & \\ \vdots & \text{at least one application of} \\ \vdots & \quad \text{a rule other than Thin} \\ \mathsf{F} \vdash_{\mathsf{V}} Z & \end{array}$$

and $\mathsf{E} \supseteq \mathsf{F}$ and Z identifies a maximal fixed point formula, $\nu Z.\, \Phi$, and for any other fixed point variable Y on this path, Z subsumes Y

Unsuccessful terminal $\mathsf{F} \vdash_{\mathsf{V}} \Psi$

1. $\Psi = \texttt{ff}$ or $\Psi = Z$ and Z is free in the initial formula and $\mathsf{F} \not\subseteq \mathsf{V}(Z)$
2. $\Psi = \langle K \rangle \Phi$ and for some $F \in \mathsf{F}$, $K(\{F\}) = \emptyset$
3. $\Psi = Z$ and there is a goal $\mathsf{E} \vdash_{\mathsf{V}} Z$ above $\mathsf{F} \vdash_{\mathsf{V}} Z$, such that there is a path as follows

$$\begin{array}{ll} \mathsf{E} \vdash_{\mathsf{V}} Z & \\ \quad\vdots & \text{at least one application of} \\ \quad\vdots & \text{a rule other than Thin} \\ \mathsf{F} \vdash_{\mathsf{V}} Z & \end{array}$$

and $\mathsf{E} \subseteq \mathsf{F}$ and Z identifies a minimal fixed point formula, $\mu Z.\, \Phi$, and for any other fixed point variable Y on this path, Z subsumes Y

FIGURE 7.3. Terminal goals in a tableau

cycles through the terminal goal $\mathsf{F} \vdash_{\mathsf{V}} Z$ (with the play repeatedly jumping to the companion goal) is won by the refuter.

A successful tableau for $\mathsf{E} \vdash_{\mathsf{V}} \Phi$ is a finite proof tree, all of whose terminal goals are successful. A successful tableau only contains true goals. The proof of this is deferred until the next section.

Proposition 1 *If* $\mathsf{E} \vdash_{\mathsf{V}} \Phi$ *has a successful tableau, then* $\mathsf{E} \models_{\mathsf{V}} \Phi$.

However, as the proof system stands, the converse is not true. A further termination condition is needed. But this condition is a little complex, so we defer its discussion until the next section. Instead, we present various examples that can be proved without it. Below we often write $E \vdash_{\mathsf{V}} \Phi$ instead of $\{E\} \vdash_{\mathsf{V}} \Phi$, and also drop the index V when it is not germane to the proof.

Example 1 The model checking game showing that $\texttt{Cnt}$ has the property $\nu Z.\, \langle \mathrm{up} \rangle Z$ consists of an infinite set of positions. $\texttt{Cnt}$ is the infinite state process, $\texttt{Cnt} \stackrel{\mathrm{def}}{=} \mathrm{up}.(\texttt{Cnt} \mid \mathrm{down}.0)$. There is a very simple proof of the property using tableaux. Let $\texttt{Cnt}_0$ be $\texttt{Cnt}$ and let $\texttt{Cnt}_{i+1}$ be $\texttt{Cnt}_i \mid \mathrm{down}.0$ for $i \geq 0$.

$$\dfrac{\dfrac{\dfrac{\dfrac{\texttt{Cnt} \vdash \nu Z.\, \langle \mathrm{up} \rangle Z}{\{\texttt{Cnt}_i \,:\, i \geq 0\} \vdash \nu Z.\, \langle \mathrm{up} \rangle Z}}{(*)\; \{\texttt{Cnt}_i \,:\, i \geq 0\} \vdash Z}}{\{\texttt{Cnt}_i \,:\, i \geq 0\} \vdash \langle \mathrm{up} \rangle Z}}{\{\texttt{Cnt}_i \,:\, i \geq 1\} \vdash Z}$$

Thin is used immediately to extend the set of processes from $\texttt{Cnt}$ to $\texttt{Cnt}_i$ for all $i \geq 0$. The application of the $\langle \mathrm{up} \rangle$ rule employs the choice function f, which we have left implicit in the proof and which maps each $\texttt{Cnt}_i$ to $\texttt{Cnt}_{i+1}$. The final goal is a successful terminal, by condition 3 of Figure 7.3, owing to the goal $(*)$ above it.

The tableau proof system also applies to finite state processes, as illustrated in the next example.

Example 2 We consider the process $\texttt{D}$ described in Figure 5.1 and the following property Φ:

$$\nu Z.\, \mu Y.\, [a]((\langle b \rangle \texttt{tt} \wedge Z) \vee Y).$$

Process $\texttt{D}$ has the property Φ. A successful tableau proving this is given in Figure 7.4. The terminal goals are all successful. The goal $(**)$ is terminal because of its companion $(*)$. Although both variables Z and Y occur on the path between these goals, Z subsumes Y and Z identifies a maximal fixed point formula; therefore, the terminal goal $(**)$ is successful. Notice that the goal $\mathbf{2} : \texttt{D} \vdash Y$ is not a leaf, even though there is the goal $\mathbf{1} : \texttt{D} \vdash Y$ above it: this is because Z occurs on the path between these goals, at $(*)$, and Y does not subsume Z.

Example 2 illustrates an application of the tableau proof system to a finite state example. However, in this case the proof is really very close to the game. Of more interest is that we can use the proof system to provide a much more succinct presentation of a verifier's winning strategy by considering sets of processes.

Example 3 Recall the level crossing of Figure 1.10. Its safety property is expressed as $\nu Z.([\overline{\texttt{tcross}}]\texttt{ff} \vee [\overline{\texttt{ccross}}]\texttt{ff}) \wedge [-]Z$. Let Φ be this formula. We employ the abbreviations in Figure 1.12, and let E be the full set $\{\texttt{Crossing}\} \cup \{E_1, \ldots, E_{11}\}$.

$$\mathtt{D} \vdash \Phi$$
$$\mathtt{D} \vdash Z$$
$$\mathtt{D} \vdash \mu Y.\, [a]((\langle b\rangle \mathtt{tt} \wedge Z) \vee Y)$$
$$\mathbf{1} : \mathtt{D} \vdash Y$$
$$\mathtt{D} \vdash [a]((\langle b\rangle \mathtt{tt} \wedge Z) \vee Y)$$
$$\mathtt{D}' \vdash (\langle b\rangle \mathtt{tt} \wedge Z) \vee Y$$
$$\mathtt{D}' \vdash \langle b\rangle \mathtt{tt} \wedge Z$$
$$\mathtt{D}' \vdash \langle b\rangle \mathtt{tt} \qquad (*)\ \mathtt{D}' \vdash Z$$
$$\mathtt{0} \vdash \mathtt{tt} \qquad \mathtt{D}' \vdash \mu Y.\, [a]((\langle b\rangle \mathtt{tt} \wedge Z) \vee Y)$$
$$\mathtt{D}' \vdash Y$$
$$\mathtt{D}' \vdash [a]((\langle b\rangle \mathtt{tt} \wedge Z) \vee Y)$$
$$\mathtt{D} \vdash (\langle b\rangle \mathtt{tt} \wedge Z) \vee Y$$
$$\mathbf{2} : \mathtt{D} \vdash Y$$
$$\mathtt{D} \vdash [a]((\langle b\rangle \mathtt{tt} \wedge Z) \vee Y)$$
$$\mathtt{D}' \vdash (\langle b\rangle \mathtt{tt} \wedge Z) \vee Y$$
$$\mathtt{D}' \vdash \langle b\rangle \mathtt{tt} \wedge Z$$
$$\mathtt{D}' \vdash \langle b\rangle \mathtt{tt} \qquad (**)\ \mathtt{D}' \vdash Z$$
$$\mathtt{0} \vdash \mathtt{tt}$$

FIGURE 7.4. A successful tableau for $\mathtt{D} \vdash \Phi$

Below is a successful tableau showing that the crossing has this property.

$$\mathtt{Crossing} \vdash \Phi$$
$$\mathtt{E} \vdash \Phi$$
$$\mathtt{E} \vdash Z$$
$$\mathtt{E} \vdash ([\overline{\mathtt{tcross}}]\mathtt{ff} \vee [\overline{\mathtt{ccross}}]\mathtt{ff}) \wedge [-]Z$$
$$\mathtt{E} \vdash [\overline{\mathtt{tcross}}]\mathtt{ff} \vee [\overline{\mathtt{ccross}}]\mathtt{ff} \qquad \mathtt{E} \vdash [-]Z$$
$$\mathtt{E} - \{E_5, E_7\} \vdash [\overline{\mathtt{tcross}}]\mathtt{ff} \qquad \mathtt{E} - \{E_4, E_6\} \vdash [\overline{\mathtt{ccross}}]\mathtt{ff} \qquad \mathtt{E} \vdash Z$$
$$\emptyset \vdash \mathtt{ff} \qquad \emptyset \vdash \mathtt{ff}$$

Notice the essential use of the Thin rule at the first step.

Exercises

1. Recall the definitions of $[\![\ ]\!]$, $[\![K]\!]$, $\langle\ \rangle$ and $\langle K\rangle$ as fixed points in Section 5.3. Provide tableau proofs of the following

   a. $\{\mathtt{Div}_n : n \geq 0\} \models [\![\ ]\!]\,[a]\mathtt{ff}$

   b. $\mathtt{Crossing} \models [\![\mathtt{car}]\!]\,[\![\mathtt{train}]\!]\,(\langle\!\langle\overline{\mathtt{tcross}}\rangle\!\rangle\mathtt{tt} \vee \langle\!\langle\overline{\mathtt{ccross}}\rangle\!\rangle\mathtt{tt})$

   c. $\mathtt{Protocol} \models [\![\mathtt{in}(m)]\!]\,(\langle\!\langle\overline{\mathtt{out}}(m)\rangle\!\rangle\mathtt{tt} \wedge [\![\{\mathtt{in}(m) : m \in D\}]\!]\,\mathtt{ff})$,

   where $\mathtt{Div}_i$ for each i is defined in Section 2.4.

2. Prove $\{\mathtt{Sched}_n : n > 1\} \models \nu Z.\,\langle -\rangle\mathtt{tt} \wedge [-]Z$.

3. The safety property of the slot machine in Figure 1.15, that the machine never pays out more than it has in its bank, as described in example 1 of Section 5.7, is given by the formula $\nu Z.\, Q \wedge [-]Z$ relative to the valuation V that assigns the set $\mathsf{P} - \{\mathtt{SM}_j : j < 0\}$ to the variable Q. Give a successful tableau demonstrating that the slot machine $\mathtt{SM}_n$ has this property.

4. Give successful tableaux for the following

   a. $\mathtt{Cl}_1 \models \nu Z.\,\langle\mathtt{tick}\rangle\langle\mathtt{tock}\rangle Z$

   b. $\mathtt{Cl}_1 \models \mu Y.\,\langle\mathtt{tock}\rangle\mathtt{tt} \vee [-]Y$

   c. $\mathtt{T}(19) \vdash \mu Y.\,\langle -\rangle\mathtt{tt} \wedge [-\overline{\mathtt{out}}(1)]Y$

   d. $\mathtt{Sem} \mid \mathtt{Sem} \mid \mathtt{Sem} \vdash \nu Z.\,[\mathtt{get}](\mu Y.\,\langle -\rangle\mathtt{tt} \wedge [-\mathtt{put}]Y) \wedge [-]Z$

   e. $\mathtt{A}_5 \mid \mathtt{A}_5 \models \mu Z.\,\langle -\rangle\mathtt{tt} \wedge [-a]Z$,

   where $\mathtt{A}_5$ is defined in example 2 of Section 5.7.

7.4 Tableaux II

In the previous section, a tableau proof system was presented for showing that sets of processes have properties. The idea is to build proofs around goals of the form $\mathsf{E} \vdash_{\mathsf{V}} \Phi$, which, if successful, show that each process in E has the property Φ relative to V. As it stands, the proof system is not complete.

An example for which there is not a successful tableau is given by the following cell, from example 2 of Section 5.5.

$$\mathtt{C} \stackrel{\text{def}}{=} \mathtt{in}(x).\mathtt{B}_x \quad \text{where } x : \mathbb{N}$$

$$\mathtt{B}_{n+1} \stackrel{\text{def}}{=} \mathtt{down}.\mathtt{B}_n \quad \text{for } n \geq 0$$

This cell has the property of eventual termination, $\mu Z.[-]Z$. The only possible tableau for $\mathsf{C} \vdash \mu Z.[-]Z$ up to inessential applications of Thin is as follows.

$$\cfrac{\mathsf{C} \vdash \mu Z.[-]Z}{\cfrac{\mathsf{C} \vdash Z}{\cfrac{\mathsf{C} \vdash [-]Z}{\cfrac{\mathbf{1}: \{\mathsf{B}_i : i \geq 0\} \vdash Z}{\cfrac{\mathbf{2}: \{\mathsf{B}_i : i \geq 0\} \vdash [-]Z}{\mathbf{3}: \{\mathsf{B}_i : i \geq 0\} \vdash Z}}}}}$$

The goal **3** is terminal because of the repeat goal **1**, and it is unsuccessful because Z identifies a least fixed point formula. However, the verifier wins the game $\mathsf{G}(\mathsf{C}, \mu Z.[-]Z)$, as the reader can verify. One solution to the problem is to permit induction on top of the current proof system. The goal labelled **1** is provable using induction on i. There is a successful tableau for the base case $\mathsf{B}_0 \vdash Z$, and assuming successful tableaux for $\mathsf{B}_i \vdash Z$ for all $i \leq n$, it is straightforward to show that there is also one for $\mathsf{B}_{n+1} \vdash Z$. However, we wish to avoid explicit induction principles. Instead, we shall present criteria for success that capture player V's winning strategy. This requires one more condition for termination.

The additional circumstance for being a terminal goal of a proof tree concerns least fixed point variables. A goal $\mathsf{F} \vdash_{\mathsf{V}} Z$ is also a terminal if it obeys the (almost repeat) condition of Figure 7.5. This circumstance is very similar to condition 3 of Figure 7.3 of the previous section for being an unsuccessful terminal goal, except that here $\mathsf{F} \subseteq \mathsf{E}$. It is also similar to condition 3 for being a successful terminal, except that here Z identifies a least fixed point variable. Not all terminal goals that obey this new condition are successful. The definition of success (taken essentially from Bradfield and the author [11]) is intricate, and requires some notation.

New terminal $F \vdash_{\mathsf{V}} Z$

4. There is a goal $\mathsf{E} \vdash_{\mathsf{V}} Z$ above $\mathsf{F} \vdash_{\mathsf{V}} Z$ such that there is a path as follows

$\mathsf{E} \vdash_{\mathsf{V}} Z$

$\vdots$ at least one application of

$\vdots$ a rule other than Thin

$\mathsf{F} \vdash_{\mathsf{V}} Z$

and $\mathsf{E} \supseteq \mathsf{F}$ and Z identifies a minimal fixed point formula, $\mu Z.\,\Phi$, and for any other fixed point variable Y on this path, Z subsumes Y

FIGURE 7.5. New terminal goal in a tableau

A terminal goal that obeys condition 3 of Figure 7.3 for being a successful terminal, or the new terminal condition of Figure 7.5, is called a "σ-terminal," where σ may be instantiated by a ν or μ depending on the fixed point variable Z.

A node of a tableau (which is just a proof tree) contains a goal, and we use boldface numbering to indicate nodes. For instance, $\mathbf{n} : E \vdash_V \Phi$ picks out the node $\mathbf{n}$ of the tableau that has the goal $E \vdash_V \Phi$. Suppose node $\mathbf{n}'$ is an immediate successor of $\mathbf{n}$, and $\mathbf{n}$ contains the goal $\mathsf{E} \vdash_V \Phi$ and $\mathbf{n}'$ contains $\mathsf{E}' \vdash_V \Phi'$.

$$\frac{\mathbf{n} : \mathsf{E} \vdash \Phi}{\ldots \ \mathbf{n}' : \mathsf{E}' \vdash \Phi' \ \ldots}$$

The ... on both sides of the bottom goal suggest that there may be another subgoal in one of these positions. A game play proceeding through (E, Φ) where $E \in \mathsf{E}$ can have as its next configuration (E', Φ'), where $E' \in \mathsf{E}'$ provided the rule applied at $\mathbf{n}$ is not the rule Thin. Which possible processes $E' \in \mathsf{E}'$ can be in this next configuration depend on the structure of Φ. This motivates the following notion. We say that $E' \in \mathsf{E}'$ at $\mathbf{n}'$ is a "dependant" of $E \in \mathsf{E}$ at $\mathbf{n}$ if

- the rule applied to $\mathbf{n}$ is $\wedge$, $\vee$, σZ, Z or Thin, and $E = E'$, or
- the rule is $[K]$ and $E \stackrel{a}{\longrightarrow} E'$ for some $a \in K$, or
- the rule is $\langle K \rangle$, and $E' = f(E)$ where f is the choice function.

All the possibilities are covered here. An example is that each B_i at node **2** in the earlier tableau is a dependant of the same B_i at node **1**, and each B_i at **1** is a dependant of C at the node directly above **1**, since the rule applied is the $[K]$ rule, and for each B_i there is the transition $\mathsf{C} \stackrel{\mathtt{in}(i)}{\longrightarrow} \mathsf{B}_i$.

The "companion" of a σ-terminal is the most recent node above it that makes it a terminal. (There may be more than one node above a σ-terminal that makes it a terminal, hence we take the lowest.) Next, we define the notion of a "trail."

Definition 1 Assume that node $\mathbf{n_k}$ is a μ-terminal and node $\mathbf{n_1}$ is its companion. A trail from process E_1 at $\mathbf{n_1}$ to E_k at $\mathbf{n_k}$ is a sequence of pairs of nodes and processes $(\mathbf{n_1}, E_1), \ldots, (\mathbf{n_k}, E_k)$ such that for all i with $1 \leq i < k$ either

1. E_{i+1} at $\mathbf{n_{i+1}}$ is a dependant of E_i at $\mathbf{n_i}$, or
2. $\mathbf{n_i}$ is the immediate predecessor of a σ-terminal node $\mathbf{n}'$ (where $\mathbf{n}'$ is different from $\mathbf{n_k}$) whose companion is $\mathbf{n_j}$ for some $j : 1 \leq j \leq i$, and $\mathbf{n_{i+1}} = \mathbf{n_j}$ and E_{i+1} at $\mathbf{n}'$ is a dependant of E_i at $\mathbf{n_i}$.

A simple trail from B_2 at **1** to B_1 at **3** in the earlier tableau earlier is:

$$(\mathbf{1}, \mathsf{B}_2)\,(\mathbf{2}, \mathsf{B}_2)\,(\mathbf{3}, \mathsf{B}_1).$$

Here B_2 at **2** is a dependant of B_2 at **1**, and B_1 at **3** is a dependant of B_2 at **2**. Condition 2 of Definition 1 is necessary to take account of the possibility of embedded fixed points as pictured in Figure 7.6. A trail from $(\mathbf{n_1}, E_1)$ to $(\mathbf{n_k}, E_k)$ may pass through $\mathbf{n_j}$ repeatedly before continuing to $\mathbf{n_k}$ via $\mathbf{n_l}$. In this case, $\mathbf{n_k}$ is a μ-terminal, meaning Z identifies a least fixed point formula. However, $\mathbf{n_l}$ may be either a μ or a ν-terminal: in both cases $\mathsf{F}_l \subseteq \mathsf{E}_l$ and in both cases Z must subsume Y because

$$
\begin{array}{ll}
\mathbf{n_1} : \mathsf{E} \vdash Z & \\
\vdots & \\
\mathbf{n_i} : \mathsf{E}_1 \vdash Y & \\
\vdots & \quad \ddots \\
\vdots & \qquad \mathbf{n_j} : \mathsf{F}' \vdash \Phi' \\
\mathbf{n_k} : \mathsf{F} \vdash Z & \qquad \mathbf{n_l} : \mathsf{F}_1 \vdash Y
\end{array}
$$

FIGURE 7.6. Embedded terminals: $\mathsf{F} \subseteq \mathsf{E}$ and $\mathsf{F}_1 \subseteq \mathsf{E}_1$.

$\mathbf{n_k}$ labels a leaf. In fact, node $\mathbf{n_i}$ here could be $\mathbf{n_1}$ with $\mathbf{n_k}$ and $\mathbf{n_l}$ both sharing the same companion. There can be additional iterations of embedded σ-terminals. For instance, there could be another σ-companion along the path from $\mathbf{n_i}$ to $\mathbf{n_l}$ whose partner labels a σ-terminal, and so on. There is an intimate relation between the sequence of processes in a trail and a sequence of processes in part of a game play from a companion node to a σ-terminal.

The companion node $\mathbf{n}$ of a μ-terminal induces a relation $\rhd_{\mathbf{n}}$, defined as follows.

Definition 2 $E \rhd_{\mathbf{n}} F$ if there is a trail from E at $\mathbf{n}$ to F at a μ-terminal whose companion is $\mathbf{n}$.

For example, in the earlier tableau, $\mathsf{B}_2 \rhd_{\mathbf{1}} \mathsf{B}_1$ because of the above trail $(\mathbf{1}, \mathsf{B}_2)(\mathbf{2}, \mathsf{B}_2)(\mathbf{3}, \mathsf{B}_1)$. More generally, $\mathsf{B}_{i+1} \rhd_{\mathbf{1}} \mathsf{B}_i$ for any $i \geq 0$.

A ν-terminal node is successful, as described in the previous section. The definition of when a μ-terminal, $\mathbf{n}' : E \vdash_{\mathsf{V}} Z$, is successful now follows.

Definition 3 A μ-terminal whose companion is node $\mathbf{n}$ is successful if there is no infinite "descending" chain $E_0 \rhd_{\mathbf{n}} E_1 \rhd_{\mathbf{n}} \ldots$.

Success means, therefore, that the relation induced by the companion of a μ-terminal is well founded. This precludes the possibility of an infinite game play asssociated with a tableau that cycles through the node $\mathbf{n}$ infinitely often so that player R wins. If there is an infinite descending chain $E_0 \rhd_{\mathbf{n}} E_1 \rhd_{\mathbf{n}} \ldots$ then any μ-terminal whose companion is node $\mathbf{n}$ is unsuccessful.

A tableau is successful if it is finite and all its terminal goals are successful (obey either conditions 1, 2 or 3 of Figure 7.3 for being successful or is a successful μ-terminal). The tableau technique is both sound and complete for arbitrary (infinite state) processes. Again, the result is proved using the game characterization of satisfaction. The following theorem includes Proposition 1 from the previous section.

Theorem 1 *There is a successful tableau with root node* $\mathsf{E} \vdash_{\mathsf{V}} \Phi$ *iff* $\mathsf{E} \models_{\mathsf{V}} \Phi$.

Proof. Assume that the goal $\mathsf{E} \vdash_{\mathsf{V}} \Phi$ has a successful tableau. We show that player V wins every game $\mathsf{G_V}(E, \Phi)$ when $E \in \mathsf{E}$. Consider a play of such a game. The idea is to follow the play through the tableau returning to the companion node of a σ-terminal. All choices for player R are given in the tableau. Choices for player V are determined by the tableau. Suppose $(F, \Psi_1 \vee \Psi_2)$ is the position reached in the game play, and $\mathsf{F} \vdash_{\mathsf{V}} \Psi_1 \vee \Psi_2$ is the goal reached in the tableau where $F \in \mathsf{F}$ and the goal is not a leaf. Then $\mathsf{F}_1 \vdash_{\mathsf{V}} \Psi_1$ and $\mathsf{F}_2 \vdash_{\mathsf{V}} \Psi_2$ are the immediate successors in the tableau. If $F \in \mathsf{F}_1$, then Player V chooses (F, Ψ_1) as the next position in the game. If $F \notin \mathsf{F}_1$, then $F \in \mathsf{F}_2$ and player V chooses (F, Ψ_2) as the next position. Next, assume that $(F, \langle K \rangle \Psi)$ is the position reached in the game play, and $\mathsf{F} \vdash_{\mathsf{V}} \langle K \rangle \Psi$ is the goal reached in the tableau where $F \in \mathsf{F}$ and the goal is not a leaf. Then $f(\mathsf{F}) \vdash_{\mathsf{V}} \Psi$ is the immediate successor in the tableau, where f is a choice function. Player V chooses $(f(F), \Psi)$ as the next position in the game. If the game configuration is (F, Ψ) and the associated goal in the tableau is $\mathsf{F} \vdash_{\mathsf{V}} \Psi$ with Thin applied to it, then we associate the same game position with the successor goal in the tableau. Any such play must be won by player V as the reader can verify. The only interesting case to show is that a play can not proceed through a least fixed point variable Z infinitely often, where Z subsumes all other variables occurring infinitely often. This would break the well foundedness requirement on μ-terminals.

For the other half of the proof, assume that $\mathsf{E} \models_{\mathsf{V}} \Phi$. We build a successful tableau for $\mathsf{E} \vdash_{\mathsf{V}} \Phi$ by preserving truth, and with a judicious use of the Thin rule. Suppose we have built part of the tableau and goal $\mathsf{F} \vdash_{\mathsf{V}} \Psi$ still has to be developed, and $\mathsf{F} \models_{\mathsf{V}} \Psi$. If Ψ is $\mathtt{tt}$, or the variable Z that is free in the starting formula Φ, then the goal is a successful terminal. If Ψ is $\Psi_1 \wedge \Psi_2$, then the goal reduces to the subgoals $\mathsf{F} \vdash_{\mathsf{V}} \Psi_1$ and $\mathsf{F} \vdash_{\mathsf{V}} \Psi_2$. If Ψ is $\Psi_1 \vee \Psi_2$, then consider each game $\mathsf{G_V}(F, \Psi)$. By assumption, player V has a history-free winning strategy for any such game. Let F_i for $i \in \{1, 2\}$ contain the processes $F \in \mathsf{F}$ such that player V's winning strategy for $\mathsf{G_V}(F, \Psi)$ includes the rule "at (F, Ψ) choose (F, Ψ_i)." The goal $F \vdash_{\mathsf{V}} \Psi$ reduces to the subgoals $\mathsf{F}_1 \vdash_{\mathsf{V}} \Psi_1$ and $\mathsf{F}_2 \vdash_{\mathsf{V}} \Psi_2$. Clearly, $\mathsf{F}_i \models_{\mathsf{V}} \Psi_i$ and $\mathsf{F} = \mathsf{F}_1 \cup \mathsf{F}_2$. If $\Psi = [K]\Psi_1$, then the goal reduces to the subgoal $K(\mathsf{F}) \vdash \Psi_1$. If Ψ is $\langle K \rangle \Psi_1$, then again consider each game $\mathsf{G_V}(F, \langle K \rangle \Psi_1)$ for $F \in \mathsf{F}$. Player V has a history-free winning strategy for any such game. Let f be the following choice function: $f(F)$ is the process G such that "at $(F, \langle K \rangle \Psi_1)$ choose $F \overset{a}{\longrightarrow} G$" is the rule in the winning strategy for $\mathsf{G_V}(F, \Psi)$. The goal $\mathsf{F} \models_{\mathsf{V}} \Psi$ therefore reduces to the true subgoal $f(\mathsf{F}) \vdash_{\mathsf{V}} \Psi_1$. Next, suppose Ψ is $\sigma Z.\Psi_1$. Let P be the smallest transition closed set containing F as a subset. Let F_1 be the set $\| \sigma Z.\Psi_1 \|_{\mathsf{V}}^{\mathsf{P}}$. If $\mathsf{F}_1 \supseteq \mathsf{F}$ apply the rule Thin to give the subgoal $\mathsf{F}_1 \vdash_{\mathsf{V}} \sigma Z.\Psi_1$, and then one introduces the subgoal $\mathsf{F}_1 \vdash_V Z$ followed by $\mathsf{F}_1 \vdash_{\mathsf{V}} \Psi_1$. The final part of the proof is to show that the tableau is successful, and this we leave as an exercise for the reader. □

Example 1 The tableau at the start of this section, reproduced below, is now successful.

$$\dfrac{\dfrac{\dfrac{\dfrac{\dfrac{\mathtt{C} \vdash \mu Z.[-]Z}{\mathtt{C} \vdash Z}}{\mathtt{C} \vdash [-]Z}}{\mathbf{1}: \{\mathtt{B}_i : i \geq 0\} \vdash Z}}{\mathbf{2}: \{\mathtt{B}_i : i \geq 0\} \vdash [-]Z}}{\mathbf{3}: \{\mathtt{B}_i : i \geq 0\} \vdash Z}$$

The only trail from $\mathtt{B}_{i+1}$ at node **1** to node **3** is

$$(\mathbf{1}, \mathtt{B}_{i+1})\,(\mathbf{2}, \mathtt{B}_{i+1})\,(\mathbf{3}, \mathtt{B}_i)$$

and therefore $\mathtt{B}_{i+1} \rhd_1 \mathtt{B}_i$. Hence, there cannot be an infinite sequence of the form $\mathtt{B}_{j1} \rhd_1 \mathtt{B}_{j2} \rhd_1 \ldots$.

Suppose the definition of $\mathtt{B}_{i+1}$ is amended as follows.

$$\mathtt{B}_{i+1} \stackrel{\text{def}}{=} \mathtt{down}.\mathtt{B}_i + \mathtt{up}.\mathtt{B}_{i+2}$$

Each $\mathtt{B}_{i+1}$ has the extra capability of performing up. An attempt to prove that C eventually terminates yields the same tableau as above. However, the tableau is unsuccessful. There are two two trails from $\mathtt{B}_{i+1}$ at **1** to the leaf node **3**.

$$(\mathbf{1}, \mathtt{B}_{i+1})\,(\mathbf{2}, \mathtt{B}_{i+1})\,(\mathbf{3}, \mathtt{B}_i)$$
$$(\mathbf{1}, \mathtt{B}_{i+1})\,(\mathbf{2}, \mathtt{B}_{i+1})\,(\mathbf{3}, \mathtt{B}_{i+2})$$

Therefore $\mathtt{B}_{i+1} \rhd_1 \mathtt{B}_i$ and $\mathtt{B}_{i+1} \rhd_1 \mathtt{B}_{i+2}$. Consequently, there is a variety of infinite decreasing sequences from $\mathtt{B}_{i+1}$. One example is the following sequence, $\mathtt{B}_{i+1} \rhd_1 \mathtt{B}_{i+2} \rhd_1 \mathtt{B}_{i+1} \rhd_1 \ldots$.

Example 2 The crossing in Section 1.2 has the liveness property "whenever a car approaches, eventually it crosses" provided the signal is fair, as stated in Section 5.7. Let Q and R be variables and V a valuation such that Q is true when the crossing is in any state where Rail has the form $\mathtt{green}.\overline{\mathtt{tcross}}.\overline{\mathtt{red}}.\mathtt{Rail}$ (the states E_2, E_3, E_6, and E_{10} of Figure 1.12) and R holds when it is in any state where Road is $\mathtt{up}.\overline{\mathtt{ccross}}.\overline{\mathtt{down}}.\mathtt{Road}$ (the states E_1, E_3, E_7 and E_{11}). The liveness property now becomes "for any run, if Q^c is true infinitely often and R^c is also true infinitely often, then whenever a car approaches the crossing eventually it crosses," which is expressed by the following formula with free variables relative to V, $\nu Y.[\mathtt{car}]\Psi_1 \wedge [-]Y$, where Ψ_1 is

$$\mu X.\nu Y_1.(Q \vee [-\overline{\mathtt{ccross}}](\nu Y_2.(R \vee X) \wedge [-\overline{\mathtt{ccross}}]Y_2)) \wedge [-\overline{\mathtt{ccross}}]Y_1.$$

Let E be the full set $\{\mathtt{Crossing}\} \cup \{E_1, \ldots, E_{11}\}$ of processes in Figure 1.12. A proof that the crossing has this property is given as a successful tableau in stages in Figure 7.7. Throughout the tableau, we omit the index V on $\vdash_{\mathsf{V}}$.

$$\{\texttt{Crossing}\} \vdash \nu Y.[\texttt{car}]\Psi_1 \wedge [-]Y$$

$$\mathsf{E} \vdash \nu Y.[\texttt{car}]\Psi_1 \wedge [-]Y$$

$$\mathsf{E} \vdash Y$$

$$\mathsf{E} \vdash [\texttt{car}]\Psi_1 \wedge [-]Y$$

$$\mathsf{E} \vdash [\texttt{car}]\Psi_1 \qquad \mathsf{E} \vdash [-]Y$$

$$\{E_1, E_3, E_7, E_{11}\} \vdash \Psi_1 \qquad \mathsf{E} \vdash Y$$

$$\textbf{T1}$$

$$\mathsf{E}_1 = \{E_1, E_3, E_4, E_6, E_7, E_{11}\}$$

$$\textbf{T1}$$

$$\mathsf{E}_1 \vdash \Psi_1$$

$$\mathbf{1} : \mathsf{E}_1 \vdash X$$

$$\mathsf{E}_1 \vdash \nu Y_1.(Q \vee [-\overline{\texttt{ccross}}](\nu Y_2.(R \vee X) \wedge [-\overline{\texttt{ccross}}]Y_2)) \wedge [-\overline{\texttt{ccross}}]Y_1$$

$$\mathsf{E}_1 \vdash Y_1$$

$$\mathsf{E}_1 \vdash Q \vee [-\overline{\texttt{ccross}}](\nu Y_2.(R \vee X) \wedge [-\overline{\texttt{ccross}}]Y_2) \wedge [-\overline{\texttt{ccross}}]Y_1$$

$$\mathsf{E}_1 \vdash Q \vee [-\overline{\texttt{ccross}}](\nu Y_2.(R \vee X) \wedge [-\overline{\texttt{ccross}}]Y_2) \qquad \mathsf{E}_1 \vdash [-\overline{\texttt{ccross}}]Y_1$$

$$\{E_3, E_6\} \vdash Q \quad \textbf{T2} \qquad \mathsf{E}_1 \vdash Y_1$$

$$\textbf{T2}$$

$$\{E_1, E_4, E_7, E_{11}\} \vdash [-\overline{\texttt{ccross}}](\nu Y_2.(R \vee X) \wedge [-\overline{\texttt{ccross}}]Y_2)$$

$$\{E_1, E_3, E_4, E_6, E_{11}\} \vdash \nu Y_2.(R \vee X) \wedge [-\overline{\texttt{ccross}}]Y_2$$

$$\{E_1, E_3, E_4, E_6, E_7, E_{11}\} \vdash \nu Y_2.(R \vee X) \wedge [-\overline{\texttt{ccross}}]Y_2$$

$$\{E_1, E_3, E_4, E_6, E_7, E_{11}\} \vdash Y_2$$

$$\{E_1, E_3, E_4, E_6, E_7, E_{11}\} \vdash (R \vee X) \wedge [-\overline{\texttt{ccross}}]Y_2$$

$$\{E_1, E_3, E_4, E_6, E_7, E_{11}\} \vdash R \vee X \qquad \{E_1, E_3, E_4, E_6, E_7, E_{11}\} \vdash [-\overline{\texttt{ccross}}]Y_2$$

$$\{E_1, E_3, E_7, E_{11}\} \vdash R \quad \mathbf{2} : \{E_4, E_6\} \vdash X \qquad \{E_1, E_3, E_4, E_6, E_7, E_{11}\} \vdash Y_2$$

FIGURE 7.7. Successful tableau

In this tableau there is one μ-terminal, labelled **2** whose companion is **1**. The relation $\rhd_{\mathbf{1}}$ is well founded because we only have: $E_1 \rhd_{\mathbf{1}} E_4$, $E_4 \rhd_{\mathbf{1}} E_6$, and $E_3 \rhd_{\mathbf{1}} E_6$. Therefore, the tableau is successful.

Example 3 The proof that the slot machine $\mathtt{SM}_n$ of Section 1.2 has the property that a windfall can be won infinitely often is given by the following successful tableau.

$$\dfrac{\dfrac{\dfrac{\dfrac{\dfrac{\dfrac{\{\mathtt{SM}_n : n \geq 0\} \vdash \nu Y.\mu Z.\langle\overline{\mathtt{win}}(10^6)\rangle Y \vee \langle -\rangle Z}{\mathsf{E} \vdash \nu Y.\mu Z.\langle\overline{\mathtt{win}}(10^6)\rangle Y \vee \langle -\rangle Z}}{\mathsf{E} \vdash Y}}{\mathsf{E} \vdash \mu Z.\langle\overline{\mathtt{win}}(10^6)\rangle Y \vee \langle -\rangle Z}}{\mathbf{1} : \mathsf{E} \vdash Z}}{\mathbf{2} : \mathsf{E} \vdash \langle\overline{\mathtt{win}}(10^6)\rangle Y \vee \langle -\rangle Z}}{\dfrac{\mathsf{E}_1 \vdash \langle\overline{\mathtt{win}}(10^6)\rangle Y}{\mathsf{E}_1' \vdash Y} \quad \dfrac{\mathbf{3} : \mathsf{E}_2 \vdash \langle -\rangle Z}{\mathbf{4} : \mathsf{E}_2' \vdash Z}}$$

E is the set of all derivatives. The vital rules in this tableau are the disjunction at node **1**, where E_1 is exactly those processes capable of performing a $\overline{\mathtt{win}}(10^6)$ action, and E_2 is the remainder; and the $\langle K\rangle$ rule at node **3**, where f is defined to ensure that E_1 is eventually reached: for processes with less than 10^6 in the bank, f chooses events leading towards loss, so as to increase the amount in the bank; and for processes with more than 10^6, f chooses to $\overline{\mathtt{release}}(10^6)$. The formal proof requires partitioning E_2 into several classes, each parametrised by an integer n, and showing that while $n < 10^6$, n is strictly increasing over a cycle through the classes; then, when $n = 10^6$, f selects a successor that is not in E_2, so a trail from E_0 through nodes **1**, **2**, **3**, **4** terminates, and therefore node **4** is a successful μ-terminal.

Example 4 Consider the processes $\mathtt{T}(i)$ for $i \geq 1$ from Section 1.1.

$$\mathtt{T}(i) \stackrel{\text{def}}{=} \textbf{if } \mathit{even}(i) \textbf{ then } \overline{\mathtt{out}}(i).\mathtt{T}(i/2) \textbf{ else } \overline{\mathtt{out}}(i).\mathtt{T}((3i+1)/2)$$

If $\mathtt{T}(n)$ for all $n \geq 1$ stabilizes into the following cycle

$$\mathtt{T}(2) \xrightarrow{\overline{\mathtt{out}}(2)} \mathtt{T}(1) \xrightarrow{\overline{\mathtt{out}}(1)} \mathtt{T}(2) \xrightarrow{\overline{\mathtt{out}}(2)} \ldots,$$

then the following tableau is successful, and otherwise it is not. But which of these holds is not known!

$$\dfrac{\dfrac{\dfrac{\{\mathtt{T}(i) : i \geq 1\} \vdash \mu Y.\, \langle\overline{\mathtt{out}}(2)\rangle\mathtt{tt} \vee [-]Y}{\mathbf{1} : \{\mathtt{T}(i) : i \geq 1\} \vdash Y}}{\{\mathtt{T}(i) : i \geq 1\} \vdash \langle\overline{\mathtt{out}}(2)\rangle\mathtt{tt} \vee [-]Y}}{\dfrac{\mathtt{T}(2) \vdash \langle\overline{\mathtt{out}}(2)\rangle\mathtt{tt}}{\mathtt{T}(1) \vdash \mathtt{tt}} \quad \dfrac{\{\mathtt{T}(i) : i \geq 1 \wedge i \neq 2\} \vdash [-]Y}{\{\mathtt{T}(i) : i > 1\} \vdash Y}}$$

The problem is that we don't know whether the relation $\rhd_{\mathbf{1}}$ induced by the companion node **1** of the μ-terminal is well founded.

Exercises

1. Show that the verifier wins the game $\mathrm{G}(C, \mu Z.[-]Z)$.
2. The following processes are from Section 5.7.

$$
\begin{aligned}
\mathtt{A}_0 &\stackrel{\text{def}}{=} a.\sum\{\mathtt{A}_i \,:\, i \geq 0\} \\
\mathtt{A}_{i+1} &\stackrel{\text{def}}{=} b.\mathtt{A}_i \quad i \geq 0 \\
\mathtt{B}_0 &\stackrel{\text{def}}{=} a.\sum\{\mathtt{B}_i \,:\, i \geq 0\} + b.\sum\{\mathtt{B}_i \,:\, i \geq 0\} \\
\mathtt{B}_{i+1} &\stackrel{\text{def}}{=} b.\mathtt{B}_i \quad i \geq 0
\end{aligned}
$$

   Provide a successful tableau whose root is

$$\{\mathtt{A}_j \mid \mathtt{A}_j \,:\, j \geq 0\} \vdash \mu Y.\langle - \rangle \mathtt{tt} \wedge [-a]Y.$$

   Show that there is not a successful tableau when $\mathtt{A}$ is replaced by $\mathtt{B}$.
3. Complete the proof of Theorem 1.
4. For the various interpretations of Cycle(...) of Section 5.7 give successful tableaux for
   a. $\mathtt{Sched}_3' \models \mathrm{Cycle}(a_1\, a_2\, a_3)$
   b. $\mathtt{Sched}_3 \models \mathrm{Cycle}(a_1\, a_2\, a_3)$

   where the schedulers are from Section 1.4.

References

[1] Andersen, H., Stirling, C., and Winskel, G. (1994). A compositional proof system for the modal mu-calculus. *Procs 9th IEEE Symposium on Logic in Computer Science*, 144-153.

[2] Austry, D., and Boudol, G. (1984). Algebra de processus et synchronisation. *Theoretical Computer Science* **30**, 90-131.

[3] Baeten, J., and Weijland, W. (1990). *Process Algebra*. Cambridge University Press.

[4] Bernholtz, O., Vardi, M. and Wolper, P. (1994). An automata-theoretic approach to branching-time model checking. *Lecture Notes in Computer Science* **818**, 142-155.

[5] Bergstra, J., and Klop, J. (1989). Process theory based on bisimulation semantics. *Lecture Notes in Computer Science* **354**, 50-122.

[6] Benthem, J. van (1984). Correspondence theory. In *Handbook of Philosophical Logic*, Vol. II, ed. Gabbay, D. and Guenthner, F., 167-248. Reidel.

[7] Bloom, B., Istrail, S., and Meyer, A. (1988). Bisimulation cant be traced. In *15th Annual Symposium on the Principles of Programming Languages*, 229-239.

[8] Bradfield, J. (1992). *Verifying Temporal Properties of Systems*. Birkhauser.

[9] Bradfield, J. (1998). The modal *mu*-calculus alternation hierarchy is strict. *Theoretical Computer Science* **195**, 133-153.

[10] Bradfield, J. and Stirling, C. (1990). Verifying temporal properties of processes. *Lecture Notes in Computer Science* **458**, 115-125.

[11] Bradfield, J. and Stirling, C. (1992). Local model checking for infinite state spaces. *Theoretical Computer Science* **96**, 157-174.

[12] Clarke, E., Emerson, E., and Sistla, A. (1983). Automatic verification of finite state concurrent systems using temporal logic specifications: a practical approach. *Proc. 10th ACM Symposium on Principles of Programming Languages*, 117-126.

[13] Cleaveland, R. (1990). Tableau-based model checking in the propositional mu-calculus. *Acta Informatica* **27**, 725-747.

[14] The Concurrency Workbench, Edinburgh University, **http://www.dcs.ed.ac.uk/home/cwb/index.html**.

[15] Condon, A. (1992). The complexity of stochastic games. *Information and Computation* **96**, 203-224.

[16] De Nicola, R. and Hennessy, M. (1984). Testing equivalences for processes. *Theoretical Computer Science* **34**, 83-133.

[17] De Nicola, R. and Vaandrager, V. (1990). Three logics for branching bisimulation. *Proc. 5th IEEE Symposium on Logic in Computer Science*, 118-129.

[18] Emerson, E., and Clarke, E. (1980). Characterizing correctness properties of parallel programs using fixpoints. *Lecture Notes in Computer Science* **85**, 169-181.

[19] Emerson, E., and Jutla, C. (1991). Tree automata, mu-calculus and determinacy. In *Proc. 32nd IEEE Foundations of Computer Science*, 368-377.

[20] Emerson, E., Jutla, C., and Sistla, A. (1993). On model checking for fragments of μ-calculus. *Lecture Notes in Computer Science* **697**, 385-396.

[21] Esparza, J. (1997). Decidability of model-checking for concurrent infinite-state systems. *Acta Informatica* **34**, 85-107.

[22] Glabbeek, J. van (1990). The linear time–branching time spectrum. *Lecture Notes in Computer Science* **458**, 278-297.

[23] Glabbeek, J. van, and Weijland, W. (1989). Branching time and abstraction in bisimulation semantics. *Information Processing Letters* **89**, 613-618.

[24] Groote, J. (1993). Transition system specifications with negative premises. *Theoretical Computer Science* **118**, 263-299.

[25] Groote, J., and Vaandrager, F. (1992). Structured operational semantics and bisimulation as a congruence. *Information and Computation* **100**, 202-260.

[26] Hennessy, M. (1988). *An Algebraic Theory of Processes*. MIT Press.

[27] Hennessy, M., and Ingolfsdottir, A. (1993). Communicating processes with value-passing and assignment. *Formal Aspects of Computing* **3**, 346-366.

[28] Hennessy, M., and Lin, H. (1995). Symbolic bisimulations. *Theoretical Computer Science* **138**, 353-389.

[29] Hennessy, M., and Milner, R. (1985). Algebraic laws for nondeterminism and concurrency. *Journal of Association of Computer Machinery* **32**, 137-162.

[30] Hirshfeld, Y., and Moller, F. (1996). Decidability results in automata and process theory. *Lecture Notes in Computer Science* **1043**, 102-149.

[31] Hoare, C. (1985). *Communicating Sequential Processes*. Prentice Hall.

[32] Jonsson, B., and Parrow, J. (1993). Deciding bisimulation equivalence for a class of non-finite-state programs. *Information and Computation* **107**, 272-302.

[33] Kannellakis, P., and Smolka, S. (1990). CCS expressions, finite state processes, and three problems of equivalence. *Information and Computation* **86**, 43-68.

[34] Kozen, D. (1983). Results on the propositional mu-calculus. *Theoretical Computer Science* **27**, 333-354.

[35] Lamport, L. (1983) Specifying concurrent program modules. *ACM Transactions of Programming Language Systems* **6**, 190-222.

[36] Larsen, K. (1990). Proof systems for satisfiability in Hennessy-Milner logic with recursion. *Theoretical Computer Science* **72**, 265-288.

[37] Larsen, K., and Skou, A. (1991). Bisimulation through probabilistic testing. *Information and Computation* **94**, 1-28.

[38] Long, D., Browne, A., Clarke, E., Jha, S., and Marrero, W. (1994). An improved algorithm for the evaluation of fixpoint expressions. *Lecture Notes in Computer Science* **818**, 338-350.

[39] Mader, A. (1997). *Verification of modal properties using boolean equation systems.* Doctoral thesis, Technical University of Munich.

[40] Manna, Z., and Pnueli, A. (1991). *The Temporal Logic of Reactive and Concurrent Systems*. Springer.

[41] McNaughton, R. (1993). Infinite games played on finite graphs. *Annals of Pure and Applied Logic* **65**, 149-184.

[42] Milner, R. (1980). *A Calculus of Communicating Systems*. Lecture Notes in Computer Science **92**.

[43] Milner, R. (1983). Calculi for synchrony and asynchrony. *Theoretical Computer Science* **25**, 267-310.

[44] Milner, R. (1989). *Communication and Concurrency*. Prentice Hall.

[45] Milner, R., Parrow, J., and Walker, D. (1992). A calculus of mobile processes, Parts I and II. *Information and Computation* **100**, 1-77.

[46] Niwinski, D. (1997). Fixed point characterization of infinite behavior of finite state systems. *Theoretical Computer Science* **189**, 1-69.

[47] Park, D. (1981). Concurrency and automata on infinite sequences. *Lecture Notes in Computer Science* **154**, 561-572.

[48] Papadimitriou, C. (1994). *Computational Complexity*. Addison-Wesley.

[49] Plotkin, G. (1981). A structural approach to operational semantics. *Technical Report*, DAIMI FN-19, Aarhus University.

[50] Pratt, V. (1982). A decidable mu-calculus. *22nd IEEE Symposium on Foundations of Computer Science*, 421-427.

[51] Simone, R. de (1985). Higher-level synchronizing devices in Meije-SCCS. *Theoretical Computer Science* **37**, 245-267.

[52] Sistla, P., Clarke, E., Francez, N., and Meyer, A. (1984). Can message buffers be axiomatized in linear temporal logic? *Information and Control* **68**, 88-112.

[53] Stirling, C. (1987). Modal logics for communicating systems, *Theoretical Computer Science* **49**, 311-347.

[54] Stirling, C. (1995). Local model checking games. *Lecture Notes in Computer Science* **962**, 1-11.

[55] Stirling, C., and Walker, D. (1991). Local model checking in the modal mu-calculus. *Theoretical Computer Science* **89**, 161-177.

[56] Streett, R., and Emerson, E. (1989). An automata theoretic decision procedure for the propositional mu-calculus. *Information and Computation* **81**, 249-264.

[57] Taubner, D. (1989). *Finite Representations of CCS and TCSP Programs by Automata and Petri Nets*. Lecture Notes in Computer Science **369**.

[58] Thomas, W. (1995). On the synthesis of strategies in infinite games. *Lecture Notes in Computer Science* **900**, 1-13.

[59] Vardi, M., and Wolper, P. (1986). Automata-theoretic techniques for modal logics of programs. *Journal of Computer System Science* **32**, 183-221.

[60] Walker, D. (1987). Introduction to a calculus of communicating systems. *Technical Report*, ECS-LFCS-87-22, Dept. of Computer Science, Edinburgh University.

[61] Walker, D. (1990). Bisimulations and divergence. *Information and Computation* **85**, 202-241.

[62] Winskel, G. (1988). A category of labelled Petri Nets and compositional proof system. *Procs 3rd IEEE Symposium on Logic in Computer Science*, 142-153.

Index

TEXTS IN COMPUTER SCIENCE

(continued from page ii)

Merritt and Stix, Migrating from Pascal to C++

Munakata, Fundamentals of the New Artificial Intelligence

Nerode and Shore, Logic for Applications, Second Edition

Pearce, Programming and Meta-Programming in Scheme

Peled, Software Reliability Methods

Schneider, On Concurrent Programming

Smith, A Recursive Introduction to the Theory of Computation

Socher-Ambrosius and Johann, Deduction Systems

Stirling, Modal and Temporal Properties of Processes

Zeigler, Objects and Systems